The Role of *Centrosema, Desmodium,* and *Stylosanthes* in Improving Tropical Pastures

Other Titles in This Series

Azolla as a Green Manure: Use and Management in Crop Production, Thomas A. Lumpkin and Donald L. Plucknett

Irrigated Rice Production Systems: Design Procedures, Jaw-Kai Wang and Ross E. Hagan

Managing Pastures and Cattle Under Coconuts, Donald L. Plucknett

Small-Scale Processing and Storage of Tropical Root Crops, edited by Donald L. Plucknett

The Mineralogy, Chemistry, and Physics of Tropical Soils with Variable Charge Clays, Goro Uehara and Gavin Gillman

Also of Interest

Tomatoes in the Tropics, Ruben L. Villareal

World Soybean Research Conference II: Proceedings, edited by Frederick T. Corbin

*Science, Agriculture, and the Politics of Research, Lawrence Busch and William B. Lacy

Developing Strategies for Rangeland Management, National Research Council

Proceedings of the Fourteenth International Grassland Congress, edited by J. Allan Smith and Virgil W. Hays

Crop Reactions to Water and Temperature Stresses in Humid, Temperate Climates, edited by C. David Raper, Jr., and Paul J. Kramer

Third International Symposium on Pre-Harvest Sprouting in Cereals, edited by James E. Kruger and Donald E. LaBerge

Cassava, James H. Cock

Zeo-Agriculture: The Use of Natural Zeolites in Agriculture and Aquaculture, edited by Wilson G. Pond and Frederick A. Mumpton

*Available in hardcover and paperback.

Westview Tropical Agriculture Series
Donald L. Plucknett, Series Editor

The Role of Centrosema, Desmodium, *and* Stylosanthes *in Improving Tropical Pastures*
edited by R. L. Burt, P. P. Rotar, J. L. Walker, and M. W. Silvey

This integrated collection describes the importance of forage
legumes for pasture development and improvement in the tropics and
subtropics. Leading agronomists review the magnitude of the need for
pasture improvement; tropical and subtropical soil and climate environments; reports of the successful use of legumes in pasture development in a wet and a dry tropical environment; and the scope of the
problem in terms of area to be developed and development logistics
required. Three legume genera, Centrosema, Desmodium, and Stylosanthes,
are discussed in detail--information is presented on taxonomy, adaptation, distribution, productivity, and usefulness--and considerable
emphasis is placed on Rhizobium germplasm resources for these genera.
A concluding section of technical essays addresses special considerations in using tropical legumes in pasture development and presents
a coordinated multidisciplinary approach to legume exploration and
evaluation.

Dr. Burt is senior research officer, CSIRO (Commonwealth Scien-
tific and Industrial Research Organization), Division of Tropical
Crops and Pastures, at Davies Laboratory in Queensland, Australia.
Dr. Rotar is an agronomist and chairman of the Department of Agronomy
and Soil Science, College of Tropical Agriculture, University of
Hawaii. Dr. Walker is associate director for research, Science and
Technology Bureau, Office of Agriculture, U.S. Agency for International
Development, Washington, D.C. Dr. Silvey is divisional liaison officer,
CSIRO, Cunningham Laboratory in Queensland, Australia.

The Role of *Centrosema, Desmodium,* and *Stylosanthes* in Improving Tropical Pastures

edited by R. L. Burt, P. P. Rotar,
J. L. Walker, and M. W. Silvey

Routledge
Taylor & Francis Group
LONDON AND NEW YORK

First published 1983 by Westview Press, Inc.

Published 2019 by Routledge
52 Vanderbilt Avenue, New York, NY 10017
2 Park Square, Milton Park, Abingdon, Oxon OX14 4RN

Routledge is an imprint of the Taylor & Francis Group, an informa business

Copyright © 1983 Taylor & Francis

Library of Congress Catalog Card Number 83-60845

ISBN 13: 978-0-367-29562-2 (hbk)
ISBN 13: 978-0-367-31108-7 (pbk)

Contents

ix

Tables and Figures

<u>Figures</u>

Contributors

M. Asghar, Lecturer in Soil Science, School of Agriculture, University of the South Pacific, Alafua, Western Samoa

R.L. Burt, Principal Research Scientist, CSIRO, Davies Laboratory, Townsville, Queensland, Australia

D.F. Cameron, Principal Research Scientist, CSIRO, Cunningham Laboratory, Brisbane, Queensland, Australia

D.G. Cameron, Principal Agrostologist, Queensland Department of Primary Industries, Brisbane, Queensland, Australia

R.J. Clements, Principal Scientist, CSIRO, Cunningham Laboratory, Brisbane, Queensland, Australia

R.A. Date, Principal Research Scientist, CSIRO, Cunningham Laboratory, Brisbane, Queensland, Australia

I.F. Fergus, Experimental Officer, CSIRO, Cunningham Laboratory, Brisbane, Queensland, Australia

W.R. Furtick, Agronomist, Department of Agronomy and Soil Science, College of Tropical Agriculture and Human Resources, University of Hawaii, Honolulu, Hawaii

P. Gillard, Senior Research Scientist, CSIRO, Davies Laboratory, Townsville, Queensland, Australia

B. Grof, Agrostologist, CIAT, Cali, Colombia

J.B. Hacker, Principal Scientist, CSIRO, Cunningham Laboratory, Brisbane, Queensland, Australia

E.F. Henzell, Chief, CSIRO, Cunningham Laboratory, Brisbane, Queensland, Australia

B.C. Imrie, Principal Research Scientist, CSIRO, Cunningham Laboratory, Brisbane, Queensland, Australia

R.M. Jones, Principal Research Scientist, CSIRO, Cunningham Laboratory, Brisbane, Queensland, Australia

Y. Kanehiro, Soil Scientist, Department of Agronomy and Soil Science, University of Hawaii, Honolulu, Hawaii

P.C. Kerridge, Principal Research Scientist, CSIRO, Cunningham Laboratory, Brisbane, Queensland, Australia

J.M. Lenne, Pathologist, CIAT, Colombia

D.A. Little, Principal Research Scientist, CSIRO, Cunningham Laboratory, Brisbane, Queensland, Australia

C.H. Middleton, Senior Agrostologist, Queensland Department of Primary Industries, Rockhampton, Queensland, Australia

R.M. Murray, Lecturer, Department of Tropical Veterinary Science, James Cook University of North Queensland, Townsville, Queensland

D.L. Plucknett, Consultative Group, International Agricultural Research, The World Bank, Washington, D.C.

P.J. Robinson, formerly Senior Research Scientist, CSIRO, Davies Laboratory, Townsville, Queensland, Australia

P.P. Rotar, Agronomist, Department of Agronomy and Soil Science, College of Tropical Agriculture and Human Resources, University of Hawaii, Honolulu, Hawaii

M.W. Silvey, Divisional Liaison Officer, CSIRO, Cunningham Laboratory, Brisbane, Queensland, Australia.

J.K. Teitzel, Senior Agrostologist, Queensland Department of Primary Industries, South Johnstone, Queensland, Australia

L.'t Mannetje, Principal Research Scientist, CSIRO, Cunningham Laboratory, Brisbane, Queensland, Australia

I. Vallis, Senior Research Scientist, CSIRO, Cunningham Laboratory, Brisbane, Queensland, Australia

J.L. Walker, Associate Director, Research, Office of Agriculture, Bureau for Science and Technology, United States International Development Cooperation Agency, Agency for International Development, Washington, D.C.

R.J. Williams, Principal Research Scientist, CSIRO, Cunningham Laboratory, Brisbane, Queensland, Australia

W.T. Williams, Chief Research Scientist, CSIRO, Davies Laboratory, Townsville, Queensland, Australia

Foreword

Increasing population and the expanding need for food in the world are placing special responsibilities on agricultural scientists and technicians. Not only must new information and technology be generated but also existing information must be gathered, evaluated and placed in context for its potential and actual usefulness; indeed, without the latter, it may be difficult to select sensible priority areas for the development of new technology. The world need for accurate, useful information that is packaged in a way readily usable by others is great, and the gathering and evaluation of such information should receive high priority for international cooperative effort.

The study presented here is an outstanding example of international technical cooperation. The cooperating institutes were the Australian CSIRO Division of Tropical Crops and Pastures and the University of Hawaii College of Tropical Agriculture and Human Resources, Department of Agronomy and Soil Science. Both are experienced, recognized centers for tropical pastures and tropical agriculture. Other prominent scientists, experts in their own fields, also contributed their knowledge in this publication. Funding for much of the joint study was provided by the Agency for International Development under the 211(d) program.

The final product of this exemplary collaborative effort is a comprehensive but focused overview of the current status and potential use of <u>Centrosema</u>, <u>Desmodium</u>, and <u>Stylosanthes</u> species in tropical pasture systems. This information is further to suggest areas where our knowledge is most limited and further research effort clearly warranted. It is the first document of its kind and it should find a wide application as a major source of information for those interested in food production in tropical areas.

Donald L. Plucknett   William R. Furtick
Washington, D.C.   Honolulu, Hawaii

Interpretive Summary

R. L. Burt
P. P. Rotar

INTRODUCTION

In recent years a great deal has been written about malnutrition in third world countries--its undesirability and possible ways of alleviating the problem. Understandably, emphasis has been placed on ways of improving crop production since crops can be consumed by man directly. Energy conversion of renewable resources to human food via the animal in the form of milk and meat is considered less efficient. Despite major technological advances in crop production there is still an important role for tropical pastures based on the forage legume, especially for developing countries.

Legumes are a group of plants that are able to fix atmospheric nitrogen, which in time may become available to the host plant, companion grasses, the soil organic matter, and to the grazing animal. Pasture legumes are used to fulfill a variety of roles. As a forage they can be used to increase animal production. This is especially important in areas with soils of low fertility where cropping is not feasible. Legumes can be integrated into farming systems and used to improve soil fertility for subsequent crops; Dobereiner (1978) calculated that a successful intensive crop production program on the Cerrado regions of South America would consume a staggering 40 percent of the world's current output of nitrogenous fertilizer. In the developing countries forage legumes have been used along roadsides or in village grazing systems to provide low-cost feed for draught animals. Although the introduction of legumes is generally a relatively simple matter, there are many types of tropical environments for which adapted legumes are not yet available. It is relevant to note that most third world countries occur in the tropics.

There are further reasons why less emphasis has been placed upon the development and use of forage legumes. Prior to the world energy crisis and the concomitant rises in fertilizer costs, there was a tendency to hide our ignorance behind a bag of fertilizer; it is easier to increase animal production by applying fertilizer to grass than to develop grass/legume combinations. There is also a novelty factor; many of the advantages of systems based upon tropical pasture legumes were not documented until the mid-1960's and, indeed, there is still a dearth of sound research findings and experience. Finally, we are dealing with rather complex systems involving interactions between soil-pasture-animal. It is therefore more difficult to appreciate the value of grass/legume systems in terms of inputs and outputs. Not surprisingly there are many misconceptions about the potential worth of tropical legumes in pasture systems.

SCOPE OF CONTRIBUTIONS

This series of essays describes the need for, and the value of forage legumes to, pasture development and improvement in the tropics and subtropics. The essays are divided into three sections. Section I provides an overview of tropical pasture improvement, the soil and climatic environments, reports of the successful use of legumes in pasture development in a wet and a dry tropical environment, and the scope of the problem in terms of area to be developed and the logistics required to further improvement. Section II deals with three legume genera chosen for their actual and potential contribution to pasture improvement. These genera are discussed in detail--taxonomy, adaptation, distribution, productivity and usefulness as pasture plants. It is the general feeling of the authors that the pasture potential of the three genera has hardly been touched. Each of the three genera contain many more species than have been evaluated for their contribution to pasture development. Included in Section II is a statement about the Rhizobium resources that must be evaluated concurrently with their associated legumes. Any deviation from concurrent evaluation with associated legumes too frequently leads to disappointment and to the rejection of the legume as a potential contributor to pastures. Considerable emphasis is placed on the necessity to collect legumes and associated root nodules containing Rhizobium at the same time. Section III consists of a number of technical essays related to pasture development, including soils, Rhizobium, social and economic considerations and a coordinated multidisciplinary program for legume exploration and evaluation.

The contributors have limited their discussion to tropical and subtropical pasture development in those regions of the world having four or more months of rainfall. They rarely refer to the vast arid and semi-arid regions of the tropical world. The essays have been limited to only three legume genera. It should be pointed out that in Latin America Roseveare (1948) listed over 100 legume genera and more than 400 species. The three genera--Centrosema, Desmodium, and Stylosanthes--were chosen for their wide range of adaptation and utilization in tropical pastures.

Grasslands of the tropics and subtropics with a four-month or more rainy season comprise over 1048 million hectares and have an average productivity varying, according to region, from 7 to 57 tonnes of beef per 1000 hectares. Conservative estimates are that a five-fold increase in productivity is a reasonable expectation for improved pastures. The contributors emphasize the fact that most of these lands suitable for grazing are unsuitable for intensive crop production and that the only way we can make use of this valuable resource is through the grazing animal which, par excellence, is a suitable collector and efficient convertor of grassland plants into a product suitable for human consumption. The legume has to be part of these grasslands as it provides adequate dietary protein for the animal, and its associated Rhizobium fix large amounts of nitrogen each year. Conservative estimates are that on the order of 100 to 200 kg per hectare are fixed each year by the legume/Rhizobium association. This represents from $US 60 to $US 120 of fertilizer nitrogen equivalent. Such contributions cannot be ignored or overlooked considering the high cost of producing nitrogen fertilizer and the increasing scarcity of energy for its production.

The impetus for improvement of tropical pastures is derived in a large degree from the success of the CSIRO Division of Tropical Pastures in Queensland, Australia. In areas where animals on native pasture lost considerable weight during the dry season and took from four to six years to reach market weight, and where calving percentages were well below 50 percent, the introduction of legumes into existing pastures reduced weight losses and even provided a net gain during the dry season and doubled or nearly tripled the conception rates of livestock. Such success has increased the interest in the further development of large areas of the tropical world.

The essays discuss the need for pasture legumes and provide valuable insights for their proper evaluation. They stress the need to evaluate potential legume species under environmental conditions similar to those from which these plants originated. This is emphasized by several contributors who provide examples where alkaline-soil-adapted legumes failed in acid soil environments and were discarded only to be reevaluated years later in the proper environment and are now proven to be useful pasture plants. It is necessary to know the limits of the climate and soil environments of the areas chosen for improvement. Once these are known, then, and only then, should legume collections be made in those parts of the world in which similar soil and climates occur and which are known to have legumes that might be utilized in the new areas. The contributors emphasize the fact that adequate information concerning the collecting site be obtained along with the legume and its root-nodules. This includes a good description of the site, the soil, and the climate. The root nodules must be collected to ensure that the associated Rhizobium necessary for nitrogen fixation are brought to the new regions where they do not exist. Without the proper Rhizobium many potentially useful legume introductions have been discarded in preliminary evaluation trials. It is absolutely necessary to test acid-soil-tolerant legume strains in acid soil environments and alkaline-soil-tolerant legumes in alkaline soil environments. The contributors point out the fact that there are strain differences within the same legume species. Stylosanthes hamata, for example, has strains adapted to acid soil environments as well as alkaline soil environments. The alkaline-soils-adapted strains do not grow well in acid soil environments and vice versa. Preconceived notions that these genera are specifically adapted to either wet or dry environments is erroneous and is based primarily on the adaptation of the few lines evaluated. Centrosema has been described as being well adapted only to warm-humid regions of high rainfall. This is not so. There is a tremendous amount of variability within the genus. Although Centrosema pubescens is primarily adapted to the wet tropics, it has been collected from 500-600-mm-rainfall regions on the southern margin of the Guajira Peninsula of Colombia and from areas above 1000 m altitude near Santa Cruz, Bolivia (18°S).

The contributors emphasize the need for the development of an international germplasm center where collections can be kept and catalogued according to standard procedures. They point out that national, regional, or local conditions seldom permit widespread testing and evaluation and that quite often materials that are tested locally fail to perform well and are discarded when, in fact, they may be suitable in different environments. It is difficult if not impossible for any one research center to maintain materials that are of no current practical value, hence the need of an international center to perform the function of long-term maintenance and storage. Without such a facility, we are not going to be able to meet the ever-expanding need for food in future years. There is an urgent need for well-organized, multidisciplinary collection programs to be established in the near future in order to rescue germplasm from areas that are rapidly being taken over by expanding rural and urban development. Many valuable collection sites will be overrun and germplasm lost if concerted efforts are not made to collect these materials. Proper collection and evaluation procedures must be followed. The contributors emphasize the urgent need for botanical/taxonomic evaluation in conjunction with agronomic evaluation. Botanically, these genera are not well defined and have been poorly collected; existing collections do not contain specimens of all the species that were collected and in fact are highly truncated, so much so that they do not cover the range of variation present in the species described. Consequently, agriculturalists using existing taxonomic descriptions become puzzled and are quite often confused as to the proper nomenclature; in the evaluation procedures, they often add to the existing confusion. A strong plea is made for a multidisciplinary approach for collection procedures and for taxonomic and agronomic evaluations. Such an approach requires a team effort--a team composed of taxonomists, agronomists, and ecologists--using modern computational techniques for complex analytical procedures to provide adequate interpretation of data.

In evaluating plants for pastures, primary criteria are animal performance and the ability to persist in difficult environments under a wide range of management conditions. It is recommended that accessions are not discarded if they are found wanting. Quite often they may have other valuable attributes such as disease, insect, and drought resistance that may be transferred to more desirable agronomic species through breeding programs.

Although these essays are limited in scope, we hope that sufficient interest will be stimulated to carry out intensive research programs in tropical legumes for pasture development. Tropical pasture legume development in much of the world is in its infancy (the number of released cultivars of Stylosanthes species can be counted on the fingers of both hands), whereas the number of released cultivars of a temperate legume such as white clover (Trifolium repens) has many cultivars available, which were developed for specific regions.

CONCERNS AND OBSERVATIONS

We shall not attempt to list our concerns and observations in order of priority for two reasons. First, we are to some extent dealing with an unknown situation; priority listing would thus tend to be subjective. Secondly, interested bodies will tend to have their own priorities determined by logistical commitments-- staff, equipment, and so forth. We shall simply, therefore, list areas that warrant further study, the order in which they appear being determined by the order in which the subjects arose in various sections.

Taxonomy and Description

In tropical pastures we are largely, but not exclusively, dealing with a system in which legumes from South and Central America are being mixed with African grasses and sown throughout the widely scattered tropical world. Good communication between workers in the field is essential; taxonomy provides the basic language for communication.

We have seen that there are parts of the system that are malfunctioning, and there are several reasons for this. Sometimes, as in examples mentioned above, we have the same plant material given different taxonomic names--somebody in Africa cannot find anything in the Australian literature about Stylosanthes bojeri; the name has been superseded, first by S. mucronata and then by S. fruticosa. Although relevant work is being done, he does not have access to it. There is an urgent need to revise genera that are now of tremendous importance in the tropics. In doing so, taxonomists should be aware of the needs of agronomists. Tropical species can be very diverse, containing forms adapted to a wide range of climatic and ecological conditions in addition to displaying an array of agronomic phenotypes. At least some of this information should be built into taxonomic descriptions.

Plant Collection

We have already commented about the deficiencies in this area; however, we note that information obtained at the time of collection can be useful in predicting areas in which the material might be useful.

Rhizobial Collection

Legumes need to be associated with rhizobia if they are to fix nitrogen. There are two alternative strategies to ensure this. First, select plants that function effectively with the local rhizobia; this cuts down costs and may be particularly appropriate for rangeland development. Secondly, inoculating seed before sowing provides immediate benefits although it may not be long-lasting. Rhizobia inoculation would, however, be very appropriate in a cropped situation and may be necessary in areas in which certain legumes have not been grown previously (e.g., introduction of some Centrosema spp. into Africa). Recent evidence suggests that different soils may be associated with specific types of rhizobia and so it is essential that legume collections in the field must also involve sampling the local Rhizobium.

Adaptation--General

Different legumes are adapted to different edaphic and climatic conditions; a species that grows well in one set of conditions may die in another. It is important that we carefully document the various conditions leading to better legume survival in a new environment. With this information the chances of sowing persistent legume pastures are enhanced.

If such work is to be undertaken then we need to rethink the strategies available to us. We can, for example, use information from Stylosanthes to illustrate this. In Australia, with soils in the neutral pH range both S. capitata and some forms of S. hamata fail. S. scabra and the acid-loving form of S. hamata survive and are selected for commercial distribution. In the Llanos, with soils of very low pH, S. capitata thrives and emphasis is placed on this species but the alkaline form of S. hamata performs poorly and S. scabra is decimated by stem-borers. In the West Indies, only the alkaline form of S. hamata survives on alkaline soils. Each institute, therefore, selects different material for its own purposes, which may be quite different from those of other areas. It is therefore necessary to establish relevant 'gene' pools, systems of communication (including plant descriptions), and methods of selecting the material for study. Methods for integrating the resultant data are already available.

Adaptation--Soils, Climate

As the intensive study of tropical legumes and of their introductions is relatively recent, we have very little information on their adaptive characteristics. Although, for instance, we find that Stylosanthes is regarded as being suitable for infertile conditions and Centrosema for more fertile situations, there is little documented evidence to confirm these 'facts'. Most of the information on legumes comes from agronomic or mineral nutrition experiments carried out with a narrow range of genetic material. Similarly, we have little information on the physiology of these plants.

Plant Improvement

There are two well-known complementary methods of improving plant adaptation: by plant introduction (utilizing material adapted elsewhere) and by plant breeding.

To date, plant introduction has provided all but one of the commercially available tropical legumes. Provided suitable programs are undertaken, this is a less costly, fast way of providing plants. There are many areas for which adapted plants are not yet available, and so an introduction program would be preferable before attempting plant-breeding programs. In some instances, where there are reasonable collections of the species of interest and where objectives can be clearly defined, plant introduction may reach a stage of diminishing returns.

This situation will become increasingly common as our collections grow. There is a need to prepare for this time by examining genetic relationships within some of the species and genera of interest; this knowledge will expedite required plant-breeding programs.

REFERENCES

Dobereiner, J. 1978. "Potential for nitrogen fixation in tropical legumes and grasses." Pp. 13-24 in Limitations and Potentials for Biological Nitrogen Fixation in the Tropics, edited by J. Dobereiner, R.H. Burris, A. Hollaender, A.A. Francs, C.A. Negia, and D.B. Scott. New York and London: Plenum Press.

Roseveare, G.M. 1948. The Grasslands of Latin America. Commonwealth Bureau of Pastures and Field Crops Bulletin 36.

Acknowledgments

The reports in this publication were prepared primarily by research staff of Commonwealth Scientific and Industrial Research Organization, Division of Tropical Crops and Pastures, in cooperation with the University of Hawaii, Hawaii Institute of Tropical Agriculture and Human Resources, Department of Agronomy and Soil Science.

Part of the work was funded through grants to the University of Hawaii from the United States International Development Cooperation Agency, Agency for International Development 211(d): AID/csd-2833 and 211(d): AID (DSAN-G-0100) and the Australian Meat Research Committee. The views and interpretations in this publication are those of the author(s) and should not be attributed to the Agency for International Development and the Australian Meat Research Committee or to any individual acting in their behalf.

Section I
The Improvement of
Tropical Pastures

1
Introduction

J. L. Walker

The developing countries (located largely in the tropics and subtropics) have had difficult problems in the 1970's, but some progress has been made. Per capita food production by individual countries has not greatly increased in that period; because although total food production has improved, it has not exceeded the rate of population increase. Total arable lands increased in area about 4 percent between 1970 and 1978, but total population increased about 19 percent during the same period.

Raising the agricultural output on arable lands, is indeed possible, as numerous local or regional tests on the value of improved technology and superior management have demonstrated, particularly in crop production. The additional potential for increasing total agricultural output, by bringing new grazing lands into production, and by significantly improving the feed supplying power of such grazing lands may be more effective in raising agricultural production than focusing attention only on arable lands. Improving the pastures established on all those grazing lands that are not suitable for crop production, has received little attention. This volume explores the dimensions and nature of tropical pasture improvement for the production of ruminant livestock for their meat and milk products for man's benefit.

A major factor in pasture improvement is the introduction and successful management of permanent legume species in mixed swards of adapted grasses and legumes. The total benefits of utilizing adapted legumes are still being explored, but the results of research and field experience to date may be summarized as follows: (a) legumes substantially increase total feed production, (b) legumes fix large amounts of nitrogen through the action of their root nodules, and thus may largely or completely obviate the need for nitrogen fertilizers, (c) the mixed grass-legume forage is much richer in protein than grasses alone and provides balanced nutrition to grazing livestock, (d) mixed sward is effective in progressively enhancing soil productivity, with additions to the soil organic matter and concurrently producing a more mellow soil structure, and (e) mixed sward is generally more effective than grasses grown alone in controlling rainfall runoff and in providing resistance to soil erosion.

It should be stressed that in order to obtain these important benefits from the use of permanent legumes-grass pastures we require an understanding of their growth requirements relative to the requirements of the grazing animal. The wide ranging reports of research on the permanent forage legumes that are summarized in this volume, provide a basis for such effective management. When widely implemented, improved pastures should prove an important method of increasing total agricultural productivity on land not suited for cropping. This should contribute significantly toward attaining the self-sufficiency in food production desired by every developing country.

2
The Need for Animal Protein in Human Diets

J. L. Walker
P. P. Rotar

The provision of adequate quantities of dietary proteins of proper nutritional quality for man is one of the more difficult aspects of the world food problem. Provision of adequate quantities of animal products is generally conceded to be the ideal way to improve world protein nutrition (Altschul, 1965).

The human organism does not distinguish between amino acids of plants, animal, bacterial, or synthetic origin but human diets have a significant requirement for animal proteins that cannot be met with plant proteins alone. The inadequacy of plant proteins is due to their limited content of certain essential amino acids. For human diets, cereal grains are generally deficient in the essential amino acids lysine and tryptophan. The grain legumes are generally deficient in the essential amino acids methionine and cystine. By contrast, animal proteins are richer in the amino acids present in suboptimal amounts in plant foods. It has been determined that the human diet is adequately balanced by providing about 15 percent of total protein needs from animal proteins. Fortunately, these animal proteins may come from meat, milk, eggs, or fish. Of these, meat and milk are efficiently produced by grazing livestock on pastures and rangelands.

Meeting world food needs for quality proteins is complicated by the fact that human dietary habits are difficult to change. Human diets can be adequately supplemented with properly processed oilseed meals, microbial protein, and synthetics, however, strong preferences have developed for traditional foods. Consequently, people are extremely reluctant to accept what they consider to be aesthetically inferior substitutes. On the contrary, animal products are highly desired and are readily eaten by most people throughout the world, consequently it is not realistic to ignore the potential for expanding the production of animal products to meet some of the world needs for high quality proteins. In the long run, it may be far easier to increase animal production in certain areas than to alter ingrained food preference patterns developed over many centuries.

The diet of the population of a country can be categorized generally on the source and calorie-protein ratio of food and on the incidence of Kwashiorkor, a childhood disease leading to mental retardation, caused by protein deficiency. There are basically three classes of diets: 1) those containing relatively large amounts of animal protein, 2) those in which most of the calories are derived from cereals, and 3) those in which most of the calories are derived from fats, sugars, or tubers or some combination of these foods. Peoples with the high animal protein diets show no signs of protein malnutrition, those with mainly cereal diets may show some signs of protein deficiency depending upon the quality of the cereal protein consumed and those with diets consisting primarily of fats, sugars, or tubers suffer from low level of protein intake and show well-marked signs of protein malnutrition.

One may ask the question is there a protein deficiency? Maladies or diseases caused by deficiencies in the amount or character of the protein intake are widely recognized under a variety of names:

Kwashiorkor - Central Africa
Infantile pellagra - South Africa
Fatty liver disease or sugar baby - Jamaica
M'buaki - Congo States
Nutritional dystrophy or nutritional oedema syndrome - India
Distrofia pluricarencial infantil - Latin Ameria
Culebrilla - Mexico

Calculations of average protein consumption in many areas fail to reveal any great deficiency in the quantity of protein available. Protein malnutrition can and indeed does exist in spite of calculated per capita consumption figures which indicate no deficiency. Kwashiorkor does exist as a disease entity.

The term "animal protein" is often used to mean high quality protein. It should only be used as a designation of the source of protein - the quality being dependent on the amino acid composition. Proteins from animal sources are of high quality because they contain more of certain essential amino acids than do proteins from plants. This aspect of quality is illustrated from a comparison of the relative amounts of essential amino acids in selected foods (Table 2.1). The amounts are compared on a percentage basis to those in egg which has an amino acid composition that is about the same composition as required by man.

An informative way to evaluate the problem of protein deficiency in human diets is to compare animal protein consumption and the physical quality of life in tropical countries (Figure 2.1). Animal protein figures are compiled from FAO statistics for the year 1975 and published in the FAO Production Yearbook. Animal protein includes meat, milk and cheese, but not fish protein. The figures are in grams protein/capita/day and are based on a percentage of all meat, milk, etc., produced. The percentage protein was derived from figures supplied by the University of Hawaii Food Science and Human Nutrition Department. The FAO figures, although not exactly the same as those generated by other agencies (e.g. World Bank), do produce the same distribution patterns and have the same outliers. The PQLI (Physical Quality of Life Index) is an index generated by the Overseas Development Council to measure

TABLE 2.1
Percentage of Ideal Concentration of Essential Amino Acids Observed in Typical Proteins (Using Egg as 100 Percent)

[Percentage concentration in whole egg protein]

Foodstuffs	Histi- dine	Threo- nine	Valine	Leu- cine	Iso- Leucine	Lysine	Methi- onine	Phenyl- alanine	Tryp- tophan
Beef	157	90	73	87	84	141	84	70	92
Fish muscle	124	96	86	106	105	148	100	79	109
Soybean meal, low fat	138	80	76	89	97	111	53	95	127
Whole rice	81	78	88	91	84	52	106	89	118
Whole wheat	100	67	62	78	64	44	78	91	109
Cottonseed meal	128	61	69	67	64	57	53	107	118
Whole corn	119	76	76	167	103	38	97	89	55
Peanut flour	100	57	66	79	66	57	25	88	72
Dried roast beans	104	79	78	78	89	106	62	89	73
Sesame meal	106	81	67	70	63	38	53	78	93

Report of the Presidents Science Advisory Committee. 1967. The World Food Problem. Vol. II. Report of the Panel on the World Food Supply. U.S. Gov. Printing Office, Washington, D.C. p. 315.

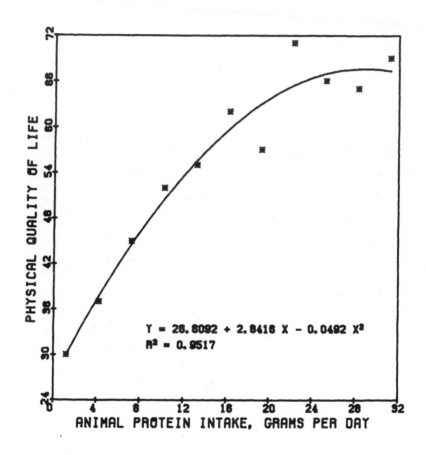

Figure 2.1. Relationship between animal protein intake and the physical quality of life for 95 developing countries in the tropics. (Each data point is the summation of several PQLI values.)

how well societies are able to satisfy certain very elemental needs of the very poor. It measures on a scale of 1 to 100, three indicators: life expectancy, infant mortality, and literacy. A composite figure, an average of the three indicator scales, is then given as the PQLI [1].

---

[1]Information for the PQLI was supplied by Mrs. Florizelle B. Liser of the Overseas Development Council. For further information see: The United States and World Development Agenda 1979, Martin M. McLaughlin, Project Director, published by Praeger Publishers, New York.

The world has about 1.2 billion cattle. Approximately 60 percent of the world's livestock are raised in developing countries; however, these countries produce only 20 percent to 30 percent of the world's meat. The low productivity of livestock in developing countries can be attributed to a number of factors, some of which have religious and cultural roots. The man reason that livestock are unproductive is lack of knowledge or of the failure to adopt scientific principles of disease control and animal husbandry. If modern principles of genetics, nutrition, range and forage management, animal husbandry, and disease control were adapted to and applied to local conditions, and farmers were appropriately supported by suppliers and processors, animal production in these countries would be increased greatly without wastefully competing with people for food.

The Republic of South Africa has developed an amazingly efficient and productive livestock industry even though the South Africans are confronted with most of the animal diseases and many of the adverse climatic factors found in the developing nations of Africa. A sizeable portion of Australia lies in the tropics. The Australians have developed excellent livestock operations in these areas despite adverse tropical conditions.

The potential for increasing world livestock production is great. Most of the major limiting livestock diseases are amenable to control. The productivity of the world's grazing lands which constitute about two-thirds of the agricultural land in the world, can be increased through improved forage, water, and range management practices. Yields of forages can be increased to much higher levels. Providing adequate diets to livestock during all seasons will result in marked increases in production.

Inadequate animal nutrition is one of the most important causes of low productivity of livestock in the developing countries. Animals, like man, have nutritional requirements which must be fulfilled. The need for proteins, fats, and vitamins for livestock vary with age, sex, rate of growth, work performance, and productivity. In primitive systems of husbandry, changes in season and weather causes changes in feed supplies which in turn cause nutritional levels to fluctuate often cyclically from adequate to inadequate. Malnutrition, if not outright starvation, is common because of deficiencies of minerals, proteins or total calories. Native vegetation without improvement or proper management and without some supplementation of required nutrients cannot sustain a productive livestock industry. The provisions of adequate quantities of essential nutrients during all seasons of the year is the key to efficient livestock production.

World food planners are often skeptical of the contribution that animals can make toward increasing world food supplies, arguing that increased animal production can be achieved only by diverting for animal use, food that otherwise would be consumed by people. Their argument is that it is wasteful to feed animals plant products such as the cereal grains that can be consumed by man because of the large loss in conversion (70-90 percent) of energy from feeds to animal products. Whether or not grains suitable for human use are fed to livestock, however, is mainly a question of the supply and demand for other uses and cost. Productive livestock systems can be developed in the absence of cereals and supplements.

REFERENCES

Altschul, A.M. 1965. Edible seed protein concentrates: their role in control of malnutrition. Israel Journal Medical Science 1:471-479.
FAO. 1975. Production Yearbook. Rome.

3
Grasslands of the Tropics

P. P. Rotar
L.'t Mannetje

Grasslands have long been used by humans for the production of animal protein either as the sole component of the diet or as a supplement to foods of vegetable origin.

Over 60 percent of the world's agricultural land is nonarable and suitable only for grazing. Animals are the only practical means of utilizing this resource for human food production. The productivity of these lands could be improved greatly if modern forage and range management techniques were utilized. Even land classified as arable, will produce under certain conditions, a more valuable human food resource through the production of high yielding forages for use as livestock feeds rather than being used to produce cereal grains for human consumption.

Substantial portions of the agricultural land in tropical or sub-tropical areas of Asia, Africa, and Latin America are grazing lands (Table 3.1). They vary greatly in topography, altitude, soil type, natural fertility, type of vegetation, available water or rainfall and climate. All types of grazing land can be improved and managed for increased production. A highly developed science and technology of range management has brought many-fold increases in the productivity of grazing lands in developed countries in temperate latitudes. The development of new and improved forages, better use of present forages, proper fertilization, water management, and weed and pest control increase the productivity of grazing lands and should be used more extensively in tropical and subtropical zones.

Little is known about many of the tropical grasses and legumes, especially about their potential as forage crops. Research is vitally needed on these forages, their productivity, chemical and nutritional composition, and on agronomic and cultural practices under various climatic conditions. The payoff in food production will be great if the results of the few research programs underway in the tropics are an accurate indication of the potential for improvement. In recent years considerable progress has been made in the development of some tropical legumes which, when mixed with tropical forage grasses, increase their carrying capacity 50 to 100 percent. Production may be further increased with the judicious use of fertilizers. With these improved pastures,

11

properly managed, the tendency for humus to disappear in a few years is markedly reduced and the soils remain productive for long periods of time. Thus the utilization of such tropical soils for livestock production results in the production of a valuable food resource as well as preservation of the soils. This is in contrast to the deterioration that follows currently practiced cultivation techniques.

Grasslands as a natural vegetation type are restricted to regions where trees do not grow because it is too dry, too wet, or too cold. In dry areas they are often associated with fertile soils and many such grasslands have since been converted to croplands. This has happened in North America, South America, the steppes of Asia, and the open downs of Australia. There is an increasing trend for grasslands to be restricted to non-arable land because of population pressures.

The tropics are geographically defined as the areas between the tropics of Cancer and Capricorn, but when the subtropics are included, this is generally extended to 30°N and S latitude; however, it is more appropriate to delimit the tropics and subtropics according to climate. This has been done (Section I-6) following the classification scheme of Troll, 1966.

The areas within the relevant climatic zones with cattle numbers and beef production are listed in Table 3.1. These data were compiled in 1978, from FAO and USDA statistics (t'Mannetje, 1978). Of the total area under consideration 23 percent is grazing lands or pastures.

Pasture improvement varies from the inclusion of a legume into existing grasses with or without fertilizer to a complete replacement of the existing vegetation with improved grasses and legumes which require annual inputs of fertilizers to remain productive. The level of input chosen is dictated by the operator's economic situation and availability of inputs - seeds, fertilizer, and productive animals. On a world scale, pasture improvement covers much less than 5 percent of the grazing lands of the tropics ('t Mannetje, 1978). Stobbs (1976) reviewed beef production in the tropics and concluded that pasture improvement could increase beef production per unit area of land six-fold. The largest constraints to pasture development are lack of knowledge, capital, and motivation.

With pasture improvement on X percent of existing grazing lands, increased beef production per unit area of land of A times the present production, total beef production from both improved and unimproved grazing lands will be Y percent of the total present production according to the formula:

$$Y = 100 + (A - 1)X$$

With improvement of 25 percent of grazing lands and a six-fold increase in beef production per unit of area of land, total production after development would be 225 percent of that before development ('t Mannetje, 1978). Applying these calculations to Africa, Central and South America with a total of 864×10^6 ha grazing lands, if 25 percent (216×10^6) were improved, beef production

TABLE 3.1
Total Areas of Land and Grazing Lands, Numbers of Cattle, and Beef and Veal Produced in 1975

Areas	Total Area ha X 10⁶	Grazing Lands ha X 10⁶	Grazing Lands %	Cattle X 10⁶	Cattle Number per 1000 ha of grazing land	Beef and veal production X 1000 t	Beef and veal production Tonnes per 1000 ha grazing land	Beef and veal production Tonnes per 1000 head of cattle
Africa[1]	1750	493	28	113	230	1400	3	12.4
America---								
Southern U.S.A.[3]	62	11	18	11	908	615	57	58.0
Central America[2]	272	81	30	48	590	1100	14	23.1
South America[2]	1430	290	20	151	520	3600	12	23.9
Asia[1]	762	31	4	236	7600	700	23	3.0
Oceania---								
Northern Australia	175	142	80	8	54	250	2	32.9
Papua New Guinea[2]	46	0.1	0.2	0.1	1300	2	20	15.4
Pacific Islands[1]	9	0.5	6	0.6	1180	10	20	17.0
Total or Mean	4506	1048	23	567	540	7677	7	13.6

[1] Data compiled by 't Mannetje, 1978.
[2] Source: F.A.O. 1975 Production Yearbook.
[3] Source: U.S.D.A. Agricultural Statistics Handbook, 1976.

would be 13.6×10^6 tonnes compared to the present production of 6.1×10^6 tonnes.

These calculations do not take into account the practical feasibility and the resources required. The improvement of 100×10^6 ha of grazing lands would require 100×10^6 kg of grass seed and an equal amount of legume seed. Assuming that 200 kg ha^{-1} is required annually for maintenance, the requirement would be 20×10^6 tonnes of superphosphate for establishment and half that amount annually thereafter. Investments in labor and fuel would also be high.

The magnitude of the seed and fertilizer requirements can be put into perspective by comparing them with recent production figures. Queensland is the major producer of tropical grass seed in the world and yet in 1974 it produced less than 2×10^6 tonnes of grass and legume seed combined. The superphosphate requirement for establishing 100×10^6 ha pasture is approaching one-fourth the world's consumption of phosphate fertilizers in 1974 (FAO, 1975). Assuming that it would cost US \$60 per ha to establish improved pasture, the total cost would be US \$60 $\times 10^9$. Assuming a carrying capacity of one animal per ha, an underestimate for humid areas, the newly developed pastures would support 100×10^6 head of livestock.

These calculations indicate that a massive allocation of resources would be needed to upgrade the productivity of a significant fraction of the grasslands in the tropics. However, this should not be a deterrent to the development of testing, and adoption of the appropriate improvement programs, as the cost/benefit results become evident.

REFERENCES

FAO. 1975. Production Yearbook. Rome.

Mannetje, L.'t 1978. The role of improved pastures for beef production in the tropics. Tropical Grasslands. 12: 1-9.

Stobbs, T.H. 1976. Beef production from sown and planted pastures in the tropics. In "Beef Cattle Production in Developing Countries," (ed.) A.J. Smith. University of Edinburgh, Edinburgh.

Troll, C. 1966. Seasonal climates of the earth. The seasonal course of natural phenomena in the different climatic zones of the earth. In "World Maps of Climatology," (eds.) E. Rodenwaldt and H.V. Juratz. Springer-Verlag. Berlin-Heidelberg-New York.

4
The Unique Role of the Ruminant Animal

D. A. Little

The influence of tropical climates and soils on herbage growth is such that the forage available to grazing animals over vast areas is of low quality for much of the year. In nutritional terms, this herbage can be characterized in the following ways:

- it is high in fiber, which is resistant to digestion and therefore to utilization by the animal,
- it is low in nitrogenous constituents, from which animals derive their protein supply,
- it is low in soluble carbohydrates, which constiute a source of readily available dietary energy, and
- it is low in many essential minerals, notably phosphorus.

Huge quantities of such material are produced in the tropics, and often much of it is burnt off during the dry season. It is virtually useless to man in any direct fashion, but ruminants are able to utilize it to produce animal protein and fat, commodities of immense value to man. In this regard, therefore, ruminants occupy a virtually unchallengeable niche in the production of human food, as well as of hides, wool, and other products.

The efficiency with which many of these massive resources are used is low. The environments concerned are often fragile, and subject to radical change as animal grazing pressure is increased. For effective utilization it is essential that the functional mechanisms and relationships between soils, plants, and animals are understood, so that the efficiency of animal production can be maximized, consistent with preservation of the environment. The peculiarities of ruminant digestion constitute an integral part of this system.

RUMEN FUNCTION AND SIGNIFICANCE

Food ingested by monogastric species, such as man, passes immediately to the stomach where it is digested under the action of hydrochloric acid and the enzyme pepsin. In ruminant herbivores on the other hand, all solid food initially passes to the rumen, a large sac, where it may remain for periods of up to 48 hours or more. Many species of bacteria and protozoa live in the rumen, and these microbes actually ferment the ingested herbage. When cattle are fed relatively low quality material, it is not unusual for the contents of the rumen to account for up to 20 percent of the animal's body weight, and about 10 percent of this material is the dry matter of consumed herbage. Thus the rumen is a highly significant organ in terms of bulk, as discussed below, and also of function.

The plant constituents of greatest nutritional significance include both structural and soluble carbohydrates, nitrogenous compounds, and minerals. The rumen microflora ferment the carbohydrates to volatile fatty acids, which are used both by the animal and the microbes as sources of energy. Before the plant cell contents can be used by the microflora they must be released by breakdown of the fiber which encases them. This is assisted by mastication and by a grinding action associated with mixing of the rumen contents by muscular contraction, but is mainly achieved by the action of the enzyme cellulase on plant cellulose. This enzyme is produced by certain bacteria but not by animals, which are therefore dependent upon these bacteria for fiber utilization.

Plant protein is also broken down to a variable extent in the rumen, yielding ammonia, amino acids, and peptides; microorganisms utilize a proportion of these products in the synthesis of microbial protein during the process of multiplication. Any ammonia not taken up by bacteria is absorbed into the circulation of the animal and converted in the liver into urea, which is then excreted either in urine or saliva. The salivary urea, of course, reaches the rumen again, where it is broken down by the bacterial enzyme urease into ammonia, which is again available for microbial utilization. This series of events clearly constitutes a process whereby the animal recycles otherwise wasted nitrogen to its rumen for potential re-use.

Material flowing from the rumen contains many bacteria; indeed the processes described ensure that a large but variable proportion of the animal's protein supply is in the form of microbial protein. On reaching the abomasum (the animal's true stomach in the monogastric sense) materials are subjected to the usual acid-pepsin digestion, following which the absorption of amino acids and peptides, as well as most minerals and fats, occurs in the small intestine. Plant fragments that may have escaped microbial degradation in the rumen are not totally lost to the system as they are subjected to further such attack in the caecum. Volatile fatty acids produced and absorbed there contribute to the animal's energy supply, but there is no evidence that microbial protein is absorbed from the large intestine in other than trace amounts, and hence is of no quantitative significance to the animal. Ammonia produced in the caecum is absorbed, however, and may enter the nitrogen recycling system mentioned above.

It is the existence of these pathways of nitrogen recycling in ruminant animals that provides the basis of their peculiar success in utilizing low quality roughage, by virtue of providing to the microflora a source of nitrogen supplementary to the low level immediately available from the diet, thus increasing the potential for fiber breakdown. This recycling is associated with another adaptive trait, the secretion of copious quantities of saliva; estimates of total daily saliva outputs ranging up to almost 200 liters/day have been made for cattle, and these quantities could well account for 10 grams of urea nitrogen or more. Although this highly developed system allows greater efficiency of nitrogen usage, the absolute amounts of nitrogen available are usually insufficient to allow for the most efficient utilization of available organic matter.

The processes described are summarized diagrammatically in Figure 4.1. The relationship between the ruminant animal and its ruminal microbial population is an excellent example of symbiosis, in which the microbes and the host animal each derive benefit from the activities of the other.

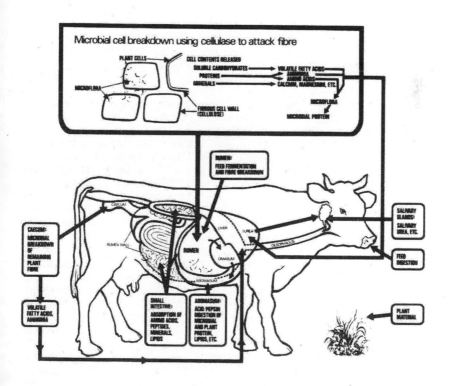

Figure 4.1. Simplified representation of some aspects of ruminant digestion.

LIMITATIONS OF NITROGEN AND LEGUME APPLICATION

In outlining ruminant digestive functions, emphasis has been placed on protein, because this is the nutrient most commonly limiting animal production in tropical areas. Other essential nutrients also frequently in short supply include phosphorus, because of very low levels of this mineral in many tropical soils, and digestible energy, because the very fibrous nature of pasture species occurring in these regions results in their being poorly digested, with a concomitantly low yield of energy. Some feeds may indeed be broken down so slowly in the rumen that 48 hours or more may elapse after ingestion before some portions pass down the gut. Clearly such delays can and do impose constraints upon the total quantity of feed that an animal physically can consume. A deficiency of any essential nutrient will limit feed intake, and, therefore, animal production.

Solving problems of low rates of animal production becomes a matter of recognizing the nutrient or nutrients responsible, and of devising methods by which the situation in the field might be alleviated. An obvious approach to overcoming the widespread problem of insufficient dietary protein is to include a legume in the pasture. Because of the relatively high levels of biologically fixed nitrogen they contain, grazing animals clearly will benefit from a diet that includes such material. Stylosanthes, Desmodium, and Centrosema have all been used with success in various tropical areas.

Another approach has been to supplement animals with a source of ammonia, usually urea; as mentioned above, the rumen microflora use this to synthesize microbial protein, which subsequently becomes available to the animal. For bacteria to do this they also need energy, phosphorus, and sulfur which are often in short supply, and the addition of molasses to the urea helps to overcome these problems. The vast majority of experimental evidence shows that the urea-molasses approach is of tangible benefit to grazing animals only when the available feed is of such poor quality that the animals are losing body weight. On the other hand, inclusion of legumes in the pasture has usually been associated with large increases in animal production during periods of weight gain, as well as increases or reduced losses during more stressful periods. These factors, in addition to problems associated with economics and availability of supplements and their handling and distribution, emphasize the vital role of pasture legumes in the improvement of animal production in the tropics.

OTHER NUTRIENT PROBLEMS OF THE TROPICS

Although the use of adapted legumes helps to overcome problems of protein shortage, it must be borne in mind that deficiencies of other nutrients may then become manifest. Indeed, in any given field situation of nutritional deficiency, it is rare for any one nutrient to be solely responsible, and the interrelationships that exist between the nutrients involved are such that the actual nutrient <u>primarily</u> responsible may vary from time to time. This introduces the concept of the first limiting nutrient, which is basic to the resolution of nutritional deficiency syndromes.

For example, as noted earlier, both nitrogen and phosphorus frequently are in short supply in tropical areas. During the wet season when pastures are actively growing, animal production responses to phosphatic supplements are often obtained, showing phosphorus to be the first limiting nutrient at that time. During the dry season, however, concentrations of crude protein in the herbage decline to levels where dietary protein becomes the first limiting nutrient. Although dietary phosphorus levels remain very low, responses to phosphatic supplements are not obtained at this stage unless the protein deficiency is first alleviated. Thus the correct identification of the first limiting nutrient, often difficult in field situations, is necessary for the application of efficient and effective supplementation programs.

The genus <u>Stylosanthes</u> is well adapted to the very low soil phosphorus conditions common in tropical areas, and it has been widely used in both sown and native pastures. Not surprisingly, the phosphorus content of these plants is relatively low, but because of biological nitrogen fixation, the crude protein content is reasonably high. Consequently, the grazing animal is presented with herbage containing a most unusually wide ratio of nitrogen to phosphorus. The important nutritional implication of this is that the presence of such a legume provides a greatly increased potential for stock to obtain a diet in which the first limiting nutrient is phosphorus. The alleviation of this problem with phosphatic supplements, combined with the protein from the legume, could greatly increase animal production in infertile areas. Many factors are involved in exploiting this situation, however, and this is an area of very active current research.

Other potential problems include, for example, the fact that legumes as a class tend to accumulate relatively small amounts of sodium. In certain circumstances a deficiency of sodium may become apparent, requiring appropriate supplementation. Soils low in sulphur have also been shown to pose problems for the grazing animal. Although sulfur levels in the plant may be adequate for its own growth, they may be too low for efficient animal production. Other mineral nutrient deficiencies, for example copper, cobalt, and zinc, are associated with low levels in soils, and the possible occurrence of these problems in any particular grazing situation should be recognized, but no further consideration of them is warranted here.

In conclusion, if the vast quantity of low quality roughage produced in tropical areas is to be utilized, ruminant animals provide a most practical means of doing so. Although ruminants are well suited to this task by virtue of their rumen function, the limitations must be recognized, and we need to work to alleviate them. Nitrogen shortage is by far the most important problem, and the inclusion in pastures of well adapted legumes is the most practical long term method of overcoming this. Phosphorus is the next most important nutrient limitation, and the correction of such deficiencies by manipulating the plant through phosphatic fertilization, or directly through the animal by the provision of supplements, has been well documented. Economic and other considerations govern the degree to which these method might be applied in any given situation. Much more research into all of these problems is required at both basic and applied levels.

5
Legumes in Action

P. P. Rotar

The family Leguminosae comprises a very large group of plants widely distributed around the world in environments of great diversity. It includes small and large herbaceous plants, viny indeterminate species, and large woody trees. It also includes many species of agricultural and economic importance, including, of course, the grain legumes that find widespread use for human food. The distinctive character of leguminous plants is that their roots may be invaded by bacteria of the genus Rhizobium which in symbiotic association with the host plant can bind or fix atmospheric nitrogen. The product of the fixation process is in part converted to bacterial protein in the nodules which develop on the root system, but more significantly, most is available to the host plant and utilized in the synthesis of protein in the leaves and later translocated to the seed. Leguminous plants in general have a higher protein content than non-leguminous plants. Legume seeds similarly have a higher protein content than those of the nutritionally important grains such as wheat or rice. It is the unique capacity of the legume to incorporate nitrogen through the fixation process to levels above that of non-leguminous plants that makes the introduction of legumes into pastures the key to the production of more nutritious forage. Particularly this is true under tropical conditions and on soils low in fertility and low in capacity for supplying nitrogen. Soil nitrogen inadequacy can be met by application of nitrogen fertilizer but economic considerations in tropical countries may make this an unattainable expedient. If legumes that are adapted to the environment can be found, the product from grass-legume pastures can be of higher protein content than the grass alone and be stable insofar as nitrogen supply is concerned.

The ideal pasture legume would probably be a long-lived perennial that persists and produces under all systems of grazing management and under all grazing pressures. It would compete well with all associated grasses, have a wide range of adaptability, and be highly productive in terms of animal performance. It would also fix, in association with Rhizobium, high amounts of nitrogen, be highly resistant to insects and diseases, and be easy to establish from seed and/or cuttings.

21

Such is, at this time, an apparent impossibility. Every tropical pasture legume tried so far has had some weakness or defect which has eliminated it from various locations or grazing situations. It is doubtful that any single legume or several combinations would be effective in all locations.

It has to be realized that the management of grass/legume pastures for optimum plant growth and performance, and the superimposition upon them of grazing animals being managed for optimum production are mutually incompatible situations. Animals grazing grass/legume pastures usually do so to the detriment of the grass/legume component of the system.

In addition to the above, a legume must be easily established under field conditions in either an existing sward or in a prepared seedbed. It is of little value to find out that although the legume establishes and performs well under experimental conditions, it is difficult if not impossible to establish and manage under field conditions. Once established it should be capable of either regenerating itself from seed or by vegetative propagation. It may be a relatively simple process in some situations to prepare a satisfactory seed bed by running a disk over dormant grass sod prior to the onset of the rainy season, and to seed the legume and have a fair chance for the legume to become established. To succesfully establish legumes into existing grass swards during the growing season, however, is a formidable task. The rapid growth rate of the grass will usually overcome young legume seedlings unless the grass is previously set back by overgrazing, strip spraying with herbicides, etc. Quite often such attempts fail with considerable loss of investment in time and money and with a total disillusionment on the part of the operator for any legume no matter how good it may be. There are no easy solutions to the problems of legume establishment in tropical pastures.

Legume persistence is related to their ability to maintain sufficient photosynthetic surface for continued productivity. Apical and axillary buds provide new stems and leaves. As these are exposed they are grazed and removed by livestock; it takes considerable time for regeneration to take place. In contrast, comparable buds in grasses are not as vulnerable to damage by the grazing animal. The gross morphology and growth habits of grasses and legumes are sufficiently different to allow the grasses to have a competitive advantage over legumes under most conditions. Hence it is very important to manage grass/legume pastures in favor of the legume.

The two examples that follow describe successful programs of pasture improvement, are from areas which are geographically close (less than 150 km apart), and are producing beef for the same market. The same socio-economic conditions therefore apply to both situations. The two areas, however, are markedly different climatically. The reader is asked to note the effect that this has on the types of pasture development undertaken and the different species selected and used.

6
The Dry Tropics—
The Kangaroo Hills Story

P. Gillard

Kangaroo Hills station is a relatively large (80,00 ha) beef cattle raising property in the monsoonal dry tropics of northern Australia. Average rainfall is 640 mm annually most of which occurs between the months of December and April; annual potential evaporation is in the vicinity of 2000 mm. Climatically this is intermediate between the V3 and V4 climates shown in Figure 6.1. The soils of the area were developed on shale and are deficient in phosphorus and nitrogen. The area is unsuitable for cropping because of the poor soils and distance from markets. Since European settlement over the last 100 years the area has supported an extensive beef cattle industry.

The native vegetation consists of an open woodland dominated by Eucalyptus crebra. Originally, the grass understory was dominated by Themeda australis which still persists in areas where access of the cattle is limited. The Themeda australis has been replaced by Heteropogon contortus in most areas which are regularly used for grazing. Although the yield of Heteropogon contortus is high, the quality of this grass declines rapidly during the dry season. This acts as a severe constraint to animal production in the region. To compensate for this poor quality, stocking rates are kept deliberately low at around one beast to 20 ha. Performance of cattle is still poor; cows do not calve every year and a branding percentage of around 60 percent is common. Bullocks may take up to five years to reach slaughter weight of 450 kg liveweight.

During the 1950s a solution to the problem of poor quality pasture was sought in the feeding of high quality legume hay. Although this was successful the hay had to be transported over very large distances at a high cost. Limited quantities of hay are still fed to selected stock in the region. Another method of improving the nutrition of the cattle during the dry season was the use of non-protein nitrogen during the 1960s. Supplements containing non-protein nitrogen have proved very useful especially for improving the calving percentage of breeding cows and are still widely used. However, supplements have not been shown to give a permanent advantage in the growth of bullocks.

Improvement of the quality of similar native pastures has

23

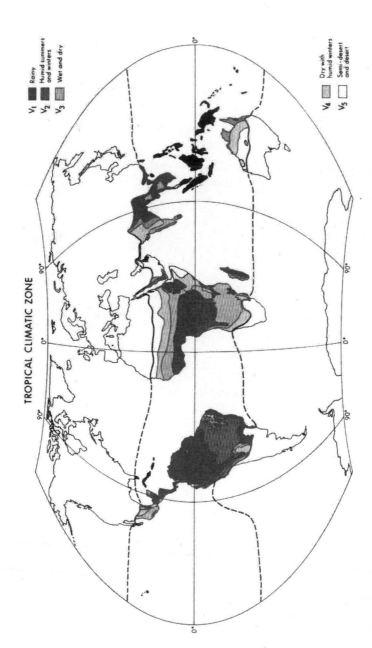

Figure 6.1. Seasonal climates of the tropics after Landsberg, et al.,1966. (V_1-wet season 9.5-12 months, V_2-wet season 7-9.5 months, V_3-wet season 4.5-7 months, V_4-wet season 0-2 months; V_4 is mar- ginal for cropping under rainfall alone, V_5 is too dry without irrigation).

already been achieved in other regions with a slightly higher rainfall (800 - 1000 mm) (Shaw, 1965; Norman and Stewart, 1964; Edye et al., 1971; Ritson et al., 1971; and Winks et al., 1974). Townsville stylo had been accidentally introduced through the Port of Townsville at the turn of the century and these experiments showed that its potential for pasture improvement depended largely on the use of superphosphate fertilizer. It had been spread widely through Northern Australia along stock routes and was known to grow in the lower rainfall regions such as Kangaroo Hills.

An experiment was established in 1965 with the object of discovering the extent of pasture improvement possible with Townsville stylo in the drier environment at Kangaroo Hills (Gillard, 1979). The experiment had treatments of tree clearing, superphosphate fertilizer and stocking rate; it covered an area of 120 ha and continued for 10 years. The results showed that Townsville stylo produced relatively poor yields in this drier environment and comprised less than 15 percent of the pasture in the best treatments. The effect of superphosphate fertilizer on both Townsville stylo yield and liveweight gain was significant only in years of above average rainfall. Tree clearing produced an increase in the total pasture yield in all years but the effect on liveweight gain was significant only in years of below average rainfall. Although the growth of Townsville stylo was poor, the cleared and fertilized treatments were able to support a much heavier stocking rate (one beast to 2.5 ha) than the native pastures (one beast to 20 ha). Despite this increased animal production, pasture improvement with Townsville stylo in the dry tropics has not been extensive. This was largely because the responses in cattle liveweight gains in most years were not large enough to compensate for the cost of the superphosphate fertilizer.

The first real opportunity for pasture improvement by a legume based technology in this dry environment occurred with the use of Stylosanthes hamata cv. Verano, an introduction from Venezuela. This species is similar to Townsville stylo but has the advantage of being able to perennate under these conditions. In small plot experiments S. hamata had given much higher yields than Townsville stylo. Another grazing experiment was therefore established at four sites (one of which was Kangaroo Hills) to compare the potential of this new introduction with that of Townsville stylo (Gillard, Edye, and Hall, 1980).

Results from this experiment showed that, under grazing, S. hamata produced the highest legume yield at all sites. At two of the sites, which had an average annual rainfall in excess of 850 mm, however, the legume yield of Townsville stylo was also high. At these sites the liveweight gains of the cattle on the S. hamata and Townsville stylo pastures was not significantly different. This suggested that a legume content of between 500 and 800 kg ha in the sward was sufficient to achieve high liveweight gains. Liveweight gains of the cattle on S. hamata pastures at Kangaroo Hills were substantially higher than on Townsville stylo pastures. The rate of gains during the wet season were higher on the S. hamata pastures and cattle also maintained weight much longer into the dry season.

Over the course of the experiment there were large changes in the composition of the S. hamata pastures at Kangaroo Hills. In the first year the legume content of the sward was 20 percent, native grasses 79 percent and Urochloa mosambicensis (a valuable grass introduced from Africa) less than 1 percent. By the third year of the experiment the legume content of the sward reached 67 percent, Urochloa mosambicensis had reached 5 percent and native grasses were reduced to 23 percent. In the sixth year soil fertility had been substantially increased by the symbiotic nitrogen fixation of the Rhizobium in the legume roots and there was a vigorous response by Urochloa mosambicensis which accounted for 52 percent of the sward; the legume content was 38 percent and native grasses 10 percent. This change in the species composition of the pasture presents a replacement of the native vegetation which is of poor quality for grazing by a high quality pasture capable of a stocking rate of one beast to 1.8 ha and a turn-off of cattle of killing weight (450 kg) after 2 years on the pasture. The major cost of this legume based pasture technology is the superphosphate fertilizer. Experiments planned for the future are designed to discover ways of obtaining greater efficiency from superphosphate fertilizer in terms of pasture and animal production.

REFERENCES

Edye, L.A., Ritson, J.B., Haydock, K.P., and Davies, J.G. 1971. Fertility and seasonal changes in liveweight of Droughtmaster cows grazing in a Townsville stylo-speargrass pasture. Australian Journal of Agricultural Research 22: 963-77.

Gillard, P. 1979. Improvement of native pasture with Townsville stylo in the dry tropics of subcoastal northern Queensland. Australian Journal of Experimental Agriculture and Animal Husbandry 19: 325-336.

Gillard, P., Edye, L.A., and Hall, R.L. 1980. Comparison of Stylosanthes humilis with S. hamata and S. subsericea in the Queensland dry tropics: Effects on pasture composition and cattle liveweight gain. Australian Journal of Agricultural Research 31: 205-220.

Landsberg, H.E., Leppman, H., Paffen, K.H., and Troll, C. 1966. In "World Maps of Climatology" (Eds.) E. Rodenweldt and H.J. Jusatz. Springer-Verlag, Berlin.

Norman, M.J.T. and Stewart, G.A. 1964. Investigations on the feeding of beef cattle in the Katherine region N.T. Journal of the Australian Institute of Agricultural Science 30: 39-46.

Ritson, J.B., Edye, L.A., and Robinson, P.J. 1971. Botanical and chemical composition of a Townsville stylo-speargrass pasture in relation to conception rate of cows. Australian Journal of Agricultural Research 22: 993-1007.

Shaw, N.H. 1961. Increased beef production from Townsville lucerne (Stylosanthes sundaica Taub.) in speargrass pastures of central coastal Queensland. Australian Journal of Experimental Agriculture and Animal Husbandry 1: 73-80.

Winks, L., Lamberth, F.C., Moir, K.W., and Pepper, Patricia M. 1974. Effects of stocking rate and fertilizer application on the performance of steers grazing stylo based pasture in North Queensland. Australian Journal of Experimental Agricultural and Animal Husbandry 14: 146-154.

7
The Wet Tropics—
The South Johnstone Story

J. K. Teitzel
C. H. Middleton

THE ENVIRONMENT

The Queensland Department of Primary Industries Research Station at South Johnstone is responsible, in Australia, for all pasture research on the wet tropical lowlands of Queensland. This fairly narrow coastal strip of land stretches from approximately 16° to 19°S latitude. Rainfalls are high, ranging from 1500 to 4000 mm with a short, but pronounced, dry season (<100 mm/month) of 3 months during the late winter and early spring. Temperatures are also high--the mean maximum and minimum at South Johnstone being 28°C and 19°C respectively. 'Low' temperatures (mean monthly minimum 13-15°C) occur in the winter dry season.

These high temperatures and rainfalls lead to high plant growth rates. Consequently there is also a high demand for land. Crops, mainly sugar and bananas, are produced on the more fertile land. The beef industry is deployed on both the poorer soils and the more fertile, but sloping, land from which potential erosion excludes crops. Many of the poor soils consist of grossly infertile residual granitic and metamorphic soils which in the past (see, for example, Sloan, Owens, and Johnstone, 1962) were considered to be useless.

We hope to show how such "useless," erosion-prone areas can be made productive by the selection of the appropriate plants and use of the appropriate technology.

PLANT INTRODUCTION AND EVALUATION

The area that we have described supported forest (dense rain forest to sparse open forest) and, in a few small areas, grassland containing Aristida, Imperata, Heteropogon, and Themeda. Even the latter are agronomically undesirable, with low yields and poor quality attributes, leading to poor animal performance.

Better pasture species were therefore required. Some species, such as common guinea grass (Panicum maximum), para grass (Brachiaria mutica), and molasses grass (Melinis minutiflora) were already available. There were, however, no native legumes which could be expected to arrest the fertility decline associated with the use of these grasses. Attempts to introduce such plants began in the late 1930s.

The early legume users had few guidelines from which to work. Temperate, sub-tropical, and tropical legumes were introduced and tested. While temperate species failed miserably, these early collections contained a few tropical legumes which were in use as 'cover' crops in the plantations in Asia. Several of these thrived, and it is from this source that centro (Centrosema pubescens), puero (Pueraria phaseoloides, Schofield stylo (Stylosanthes guianensis), and calopo (Calopogonium muconoides) were derived. Except for calopo, all are still widely used.

During this establishment period, which began in the 1930s and lasted until the 1950s, other grasses and legumes showed promise and some of these were subsequently released commercially; signal grass (Brachiaria decumbens), Hamil guinea grass (Panicum maximum) and the legume hetero (Desmodium heterophyllum) still remain as very useful plants.

By this time, associated work was already suggesting that development of the very infertile soils was feasible. It was realized that new plants might be needed for this situation and, moreover, that such plants should produce more growth than the established varieties during the colder, drier season. With these objectives, renewed interest in plant introduction and evaluation originated in the 1960s. Even with the limitations imposed, such as the inadequacy of plant material due to the lack of collecting missions, there were marked successes. Makueni guinea grass (Panicum maximum), Belalto centro (Centrosema sp. aff. pubescens and the Cook and Endeavour stylos (Stylosanthes guianensis) proved to be superior to the older cultivars in the required characteristics (see sections II-2, II-4). Additionally, reselection within the diverse 'common' guinea population within Australia led to its replacement in the late 1970s, by the cultivar Riversdate.

Plant introduction of grasses from Africa and legumes, mainly from South America, finally enabled 'useless' areas to be made productive for beef cattle. There is still scope for improvement--better plants for poorly drained soils, legumes more capable of competing with creeping grasses and plants capable of producing and persisting at lower fertilizer levels--but requirements have been identified and there are still many unexplored tropical species which are likely to provide such plants. The south Johnstone plant introduction program has been reviewed by Harding (1972).

THE IMPORTANCE OF SOILS AND FERTILIZERS

The first cattle fattening industry in the area was small and utilized the more fertile soils, experiencing relatively few pasture problems. This situation changed drastically in the 1960s when large tracts of infertile forest country were released for development. Such developments failed dismally due to low soil fertility and lack of appropriate technology responsible.

To rectify this situation there was a need, not only to

understand what the fertility problems were and how to correct them, but to communicate the results to others at a practical level. A program was initiated in the 1960s to do this.

First, the environment was surveyed and major features of the geology, soils, climate, and vegetation noted. The results were then classified to produce a working framework; a given unit carried a certain type of vegetation, had a certain soil and climate, and was formed on a given type of rock. The fertility of the soils in the units was then studied in pot experiments and the significance of the results was validated in field experiments. Subsequently, further field experiments to determine optimum fertilizer application rates under commercial conditions were carried out.

Presenting the results for practical use proved to be a problem; for instance, standard soil classification groupings did not represent mineral deficiency patterns. The best method of land classification finally proved to be one which we might intuitively have selected. A two way table is constructed in which the natural vegetation is represented in the columns and the parent rock on which it is borne in the rows. The natural vegetation is actually a good indicator of the general level of fertility and the parent rock determines which specific elements are needed; basaltic soils are low in P and Mo, and granite soils low in P, K, Ca, S, Cu, and Zn. In Table 7.1 this framework is used to show the minimum quantity of fertilizer required, as determined experimentally, to successfully establish productive pastures of commercially used cultivars on a given land unit. The species recommended are listed in Table 7.2.

Nine years of commercial results have shown that these recommendations work commercially. Additionally, because it has been possible to order a mass of complex scientific information into a simple, readily understandable system, it has been possible to communicate these results simply and reliably. This work has been reviewed by Teitzel (1979).

PASTURE DEVELOPMENT

We have already indicated that the natural vegetation of the area with which we are concerned was unsuitable for animal production and required removal. Experience at South Johnstone, under Australian economic conditions, has shown that this cannot be accomplished using the systems of slow ecological change advocated for the dry tropics (Figures 7.1 - 7.10). The aim must be to clear and quickly replace the forest with a vigorous grass/legume mixture. Failure to do so allows reversion to forest and to an unproductive pastoral situation.

Successful methods of accomplishing removal and replacement, together with procedures to be avoided, have been detailed by Teitzel, Abbott, and Mellor (1974). Briefly, the land is cleared (or plowed in the case of grasslands) at a strategic time. If plowing is done too early, regrowth from underground plant parts occurs, and if clearing is done too late the timber will not dry

TABLE 7.1
Recommended Fertilizers for Pasture Establishment

Vegetation	Parent Material				
	Basalt	Metamorphic	Granite	Mixed Alluvial	Beach Sands
Rain forest	250 P 0.5 Mo	250 P 0.5 Mo	250 P	250 P 0.5 Mo	-
Scrub	-	250 P 50 K 0.5 Mo	250 P 50 K	250 P 50 K 0.5 Mo	-
Open forest	-	500 P 100 K 0.5 Mo	500 P 50-100 K 10 Zn 10 Cu	500 P 50-100 K 0.5 Mo 10 Cu	500 P 150 K 10 Cu 10 Zn
Grassy woodland	-	-	500 P 100 K 10 Zn 10 Cu	500 P 100 K 0.5 Mo 10 Cu	250 P 10 Cu 10 Zn
Palm forest	-	250 P 0.5 Mo	250 P	250 P 0.5 Mo	-
Narrow-leaf tea-tree	-	500 P 100 K 0.5 Mo	500 P 50-100 K 10 Zn	500 P 50-100 K 0.5 Mo	N.D.
Broad-leaf tea-tree	-	500 P 100 K 0.5 Mo	500 P 100 K 10 Zn	500 P 100 K 0.5 Mo	N.D.

N.D. indicates development of these soils is not recommended
P = kg superphosphate ha^{-1}
Mo = kg sodium molybdate ha^{-1}
K = kg KCl ha^{-1}
Cu = kg CuSO ha^{-1}
Zn = kg ZnSO ha^{-1}

out sufficiently for burning. The dried material is windrowed and burnt, and seedbeds are prepared between the windrows. The preparation required depends upon the vegetation cleared. In cleared rain forest little vegetation usually remains after a burn and a single light cultivation may be all that is required whereas on more lightly timbered country or grassland, deeper plowing may be necessary to remove weedy grasses and woody understory plants.

TABLE 7.2
Grass-Legume Combinations Recommended for Various Situations

Situation	Grass[1] Legume[2] Mixture
Well-drained fertile soils	Guinea-centro puero
Well-drained soils of moderate fertility	Guinea-centro puero stylo
Well-drained soils of low fertility	Guinea-puero stylo, or Signal-puero stylo
Moderately-drained soils	Hamil-centro puero stylo
Poorly-drained soils	Para-centro puero stylo

Grasses[1]
Guinea
 grass = Panicum maximum cv. Riversdate
 Hamil = Panicum maximum cv. Hamil
 Para = Brachiaria mutica
 Signal = Brachiaria decumbens

Legumes[2]
Centro = Centrosema pubescens cv. Belalto or
         Common
Puero = Pueraria phaseoloides
Stylo = Stylosanthes guianensis cvs.
        Schofield, Cook, or Endeavor

However, care must be taken to avoid plowing too deeply and turning up the highly infertile subsoil.

Once cleared, it is necessary to establish as quickly as possible a good ground cover of the desired pasture plants. This not only avoids erosion hazards and helps to suppress the regrowth of undesirable species, but also tends to produce high yielding pastures more quickly. Large numbers of seeds are therefore sown and germination of some plants is aided by scarification of seeds. Inoculation with a Rhizobium culture is necessary to ensure that the legumes get off to a good start. If possible the seed should be drilled into the soil, although in some cases only aerial sowing may be feasible. In all cases, the seed should be covered immediately after planting either by harrowing, rolling or by a heavy fall of rain. Note the marked contrast to the less capital- and labor-intensive recommendations for extensive, dry-tropical grazing areas.

The early management of such a pasture may be critical--the pasture is competing with weeds, plants have no reserves, and,

Figure 7.1. Cattle mustered from land once described as useless. (Note the rainforest and cleared areas in the background.)

Figure 7.2. Rainforest crushed by a ball and chain pulled by two bulldozers.

Figure 7.3. A good weed free establishment population of guinea and puero resulting from proper seed bed preparation and seed distribution.

Figure 7.4. A typical response to superphosphate on an open sebrophyll forest site.

Figure 7.5. The tall vigorous pasture on the right of the fence
was fertilized with P, Cu, Zn and Mo and is carrying
3 beasts per hectare. The short pasture on the left
was fertilized with P alone and is carrying 1.5
beasts per hectare.

Figure 7.6. Loading a solution of Cu, Zn and Mo in an aircraft
for application to grass-legume pastures.

Figure 7.7.   A well-balanced, adequately fertilized guinea-centro pasture.

Figure 7.8.   Woody regrowth on a pasture which has lost its competitive vigour through inadequate maintenance fertilizer.

Figure 7.9. Normal appearance of a guinea-centro-puero pasture in a good wet season.

Figure 7.10. A guinea-centro-puero pasture at the end of an exceptionally dry season. This well-fertilized pasture easily survived such treatment.

there is no 'seed bank' in the soil to allow plants to regrow should they be damaged. With guinea grass for example, grazing should be delayed until the grass has set seed, or only light grazing practiced if the need for suppression of the associated legumes looks likely. Differences between the species should be considered. Stylo is not particularly palatable in its early stages and is sensitive to shading so early grazing of the pasture does not damage the legume. On the other hand, centro and puero are much more palatable and early grazing can kill them.

MANAGEMENT AND PRODUCTION LEVELS OF ESTABLISHED PASTURES

Ideally, we would like pasture production to parallel the needs of livestock. This, however, is not the case; in the wet season we have extremely high rates of production (>150 kg dry matter ha^{-1} day^{-1}) declining to low levels (<50 kg dry matter ha^{-1} day^{-1}) during the cooler dry season.

There are several ways to approach this problem and the method used depends very much upon socio-economic considerations. First, we can base stocking rates on the carrying capacity of the pastures during the dry season which results in inefficient utilization of the summer feed. Secondly, we can devote part of the property to the production of winter feed. Temperature-insensitive, prostrate grasses, fertilized with bag nitrogen, and located on the heavier, deeper soils (which hold moisture better in the dry season) are suitable for this purpose. Thirdly, we can use plants which are less temperature-sensitive (Belalto centro, Makueni guinea grass, and Cook stylo). In South Johnstone, a combination of the second and third strategies allows 50-100 percent higher stocking rates than the first strategy. Fourthly, (applicable to other tropical areas), we can supplement the diet of the animals during the slow growth part of the year with crop residues. Conversely, excessive summer production could be used to feed other animals or possibly production could be conserved. Finally, we can adjust stock numbers to suit the growth of the pasture by selling or removing cattle, during the cooler, drier months. Because of practical and management problems, the last two alternatives are rarely used in the Australian wet tropics. In many tropical areas, where high temperatures and rainfalls continue throughout the year, the problem of uneven feed supply does not exist.

The stocking rate imposed is determined by economic conditions, as well as by the grass/legume combination which is suited to the particular environment. Excessive stocking rates (>5 beasts/ha) during the wet season when the trailing legumes centro and puero spread and root, reduce the legume's powers of regeneration and survival. A more prostrate, strongly stoloniferous legume (hetero) is favored by heavy grazing which reduces competition from the associated grass. Stylo is sensitive to frequent and severe defoliation and also to excessive shading from vigorous grass growth. Under fertile conditions the grazing necessary to reduce the grass growth might kill it. Fortunately, stylo has a redeeming feature: it is capable of good growth at very low fertility levels. Under these conditions, excessive grass growth does not occur. Similar considerations (although less critical) apply when we consider the various associated grass species, but those will not be considered here.

With reasonable care the vigorous growth of the pasture species, together with the soil fertility maintenance which we shall mention subsequently, suppresses weed growth. Where needed weed control can be practiced, withholding stock and allowing the pasture to smother weeds may be sufficient. Mechanical slashing or chemical control can also be useful. Survival characteristics of the more important legumes to a range of chemical herbicides have been determined.

MAINTENANCE OF SOIL FERTILITY

Elsewhere in this book we have indicated the levels of nitrogen fixation which we might expect under South Johnstone, Queensland Australia environmental conditions. Absolute rates of nitrogen fixation are not considered here. The basic objective is, instead, to find the minimum mineral requirements which will provide optimum legume growth which, in turn, stimulates nitrogen fixation, stimulating grass growth. Fortunately, the levels which satisfy the legumes also meet the requirements of the associated grasses, and there has been no need to modify fertilizer applications to alter grass/legume balance. Slight tendencies to dominance can, in any case, be overcome by grazing management practices.

Present results suggest that the classification previously described (Table 7.1) may also be the most useful framework on which to base fertilizer maintenace strategies. Current work is designed to consider animal nutrition requirements, as well as the pasture and the influence of the animal on the pasture's nutrition requirements. Again these will be put into a readily communicable framework, useful at the commercial level.

LIVESTOCK PRODUCTION

Without pasture development there would not be a viable cattle industry in the area we have been describing. The native vegetation would not support cattle. The base level for comparison must be from those pastures which could be developed with some of the original grass (paragrass, molasses grass, and guinea grass) on the more fertile soils. Production levels were low (Table 7.3). Substantial application of bag nitrogen could be used to boost production as could changes in stocking rate. On a whole-property level, such practice requires large capital resources and is likely to be uneconomical as fertilizer prices rise. Inclusion of a legume in a grass dominant pasture, although not giving the high rates of production associated with bag nitrogen, has given much higher levels of animal production. On a whole-property sys-

TABLE 7.3
Some Animal Performance Data from North Queensland Lowland, Humid
Tropical Areas[1]

| | Stocking rate | Liveweight gains | |
	(beasts/ha/year)	per animal (kg/day)	per ha (kg/year)
Grass[2] alone			
(Para, molasses and guinea)[AB]	1.1	0.52	210
	2.0	0.42-0.58	305-415
Grass[2] + bag N			
Guinea + 165 kg/ha[A]	4.2	0.43	655
Signal + 196 kg/ha	4.5	0.54-0.66	870-1030
Signal + 196 kg/ha	3.5	0.57-0.62	690-740
Pangola + 225[B]	5.0	0.46	880
Grass + legume[3]			
Guinea + centro[A]	2.2-2.9	0.57-0.61	450-640
Guinea + centro[A]	2.5	0.52-0.65	475-595
Guinea + centro ± glycine[A]	3.3	0.63	760
Pangola + hetero[B]	4.1	0.52	785

[1]These results have been taken from experiments carried out on different soils and using different stocking rates. For general use the comparisons listed are sensible though the grass alone figures were recorded over 25 years ago from pastures receiving grossly inadequate fertilizer.

Compare results between pastures bearing the same capital letter.

Grasses[2]	Legumes[3]
Paragrass = <u>Brachiaria mutica</u>	centro = <u>Centrosema pubescens</u>
Molasses = <u>Melinis minutiflora</u>	glycine = <u>Glycine wightii</u>
Guinea grass = <u>Panicum maximum</u>	hetero = <u>Desmodium heterophyllum</u>
Signal grass = <u>Brachiaria decumbens</u>	
Pangola grass = <u>Digitaria decumbens</u>	

tem, integration of the winter productivity of bag nitrogen pastures with the economy of grass legume pastures is desirable in our present situation.

These results have been obtained using sub-optimal fertilizer maintenance levels. Even so, the animal production levels obtained are good for tropical pastures by world standards (Norman, 1974). Current work suggests that the use of better species and optimum fertilizer levels can give animal production levels of over 800 kg ha^{-1} yr^{-1} on rain-grown pastures.

COMMERCIAL PRODUCTION EFFICIENCY AND STABILITY

Some of the more important technological advances described above were followed by rapid, large-scale commercial development. At time of adoption, much of the technology was fragmented and when used at all, used only in isolation. Consequently, there have been a few predictable disasters in this inherently unstable ecological area. There have also been resounding commercial successes. One of the major problems is that a multitude of options are open to individual property planners, and all realistic options presently available require relatively high inputs. Unfortunately, the manager has little objective guidance in formulating the most efficiency and economical production systems for his particular set of physical, biological, and economic circumstances.

To help overcome this problem, several complementary and inter-related approaches are being used to study the productivity, efficiency and stability of a range of beef production systems covering the more important ecological units. These involve the monitoring of actual commercial produciton systems, physical modelling of alternative production systems, and computer modelling.

REFERENCES

Harding, W.A.T. 1972. The contribution of plant introduction to pasture development in the wet tropics of Queensland. Tropical Grasslands 6: 191-199.

Norman, M.J.T. 1974. Beef production from tropical pastures. Australian Meat Research Committee Review 16: 1-.

Sloan, W.J.S., Owens, A.J., and Johnstone, M.A. 1962. Report of the North Queensland Land Classification Committee, 1960-61. Office of the Minister for Lands and Irrigation.

Teitzel, J.K., Abbott, R.A. and Mellor, W. 1974. Beef cattle pastures in the wet tropics. II. Queensland Agricultural Journal 100: 149-155.

Teitzel, J.K. 1979. Formulation of pasture fertilization programs for the wet tropical coast of Australia. In "Pasture Production in Acid Soils of the Tropics." (eds.) P.A. Sanchez and L.E. Tergas, C.I.A.T. Series 03EG-5. Colombia.

8
Factors in Tropical
Pasture Improvement

R. L. Burt, J. L. Walker,
and Y. Kanehiro

CLIMATES

Throughout much of the tropics, temperatures tend to be rather high, and rainfall may be the predominant factor limiting plant growth. Often, therefore, tropical climates are described in terms of their rainfall characteristics. To ease presentation we shall initially follow this which, however, is not totally applicable. Plant distribution and growth in the tropics can be affected by temperature; indeed under other circumstances ability to withstand frost or to grow at low temperatures is deemed to be a desirable characteristic in many species.

In Figure 6.1 (Chapter 6) length of wet season (Landsberg et al., 1966) is used to discriminate between types of tropical climate. Note that three of these types have dry seasons, lasting more than 5 months with no effective rainfall. It is also clear that some continents are more favored than others; South America and much of Asia tend to have long growing seasons, whereas Africa, peninsular India, and much of Australia tend to be much drier. Australia, in fact, has only a very narrow coastal strip of the wetter types of climate (too small to appear on a world map).

To understand the significance of climate on forage availability and animal production in general, we need to study examples of more specific situations. A transect taken through West Africa passing from Equatorial Africa to Sub-Saharan regions, can be used as an example where the need to produce more animal protein is recognized (Figure 8.1).

At the southern end of the transect temperatures are consistently high, and there is little difference between day and night, winter and summer. Precipitation is high and, although there is a short dry season, moisture is stored in the soil and plant growth occurs most of the year. Moving north, however, temperatures become more extreme with higher maxima and lower minima than those found near the coast. Variations can be quite large, and there are big differences between winter and summer climates. Rainfall decreases and the length of the growing season shrinks until it is virtually non-existent. Even in those months with the highest, most reliable rainfall, there can be great variations. At Mopti,

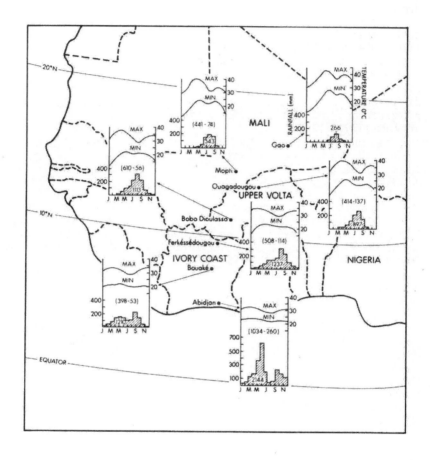

Figure 8.1.   Climatic differences in West Africa going from Equatorial Africa to the Sub-Saharan regions.

Legend:

Mean monthly maximum and minimum temperatures are plotted and average monthly rainfalls shown in a histogram.   Rainfall variability in the month with the highest rainfall is shown by numbers in brackets and total annual rainfall inside the histogram.

Mali, for instance, there have been been recorded rainfalls ranging from 74 to 441 mm in August. In these situations there is
drought or flooding.

Vegetation and land use vary dramatically with climate, and a
table taken from Whyte (1968) and modified slightly can be used to
illustrate this (Table 8.1). We note that the vegetation ranges
from desert to rainforest, and that, in the drier regions grazing
is the main or only use made of the land. Where cropping is carried out, different suites of crops are needed to cope with the
different climates, from sorghum and millet in the drier zones to
rice, bananas, and rubber in the wetter zones. Africa is the home
of virtually all of the grasses used for pastures throughout the
tropics and, for future reference, some of these are listed
against the African environments which produced them. As with the
crops, there are different suites of species for different climates. For completeness, the corresponding ranges of South American legumes are also given. South America is the region from
which most tropical pasture legumes are likely to be forthcoming.

Tropical pastures are a classical example of the necessity
for international cooperation; we sow African grasses with South
American legumes often to feed animals containing Brahman blood,
from India.

SOILS

The characteristics of soils are determined by the parent
material, the climate and the vegetation through the long soil-
forming process. A wide variety of soils occur in the tropics reflecting the diversity of the environments, contrary to some earlier beliefs in their uniformity and the omnipresence of laterite.
A taxonomic system of soil classification has been developed
(USDA, 1975) which recognizes the salient differences between
soils and can form the basis for land use and management decisions. Figures 8.2, 8.3 and 8.4 (adapted from Aubert and Tavernier, 1972 and Sanchez and Isbell, 1979) show the distribution of
soil orders through the tropical world. For the purpose of crop
production and agricultural land-use in the tropics the following
six orders are of primary interest:

Oxisols	Alfisols	Vertisols
Ultisols	Entisols	Inceptisols

The remaining four orders (Aridisols, Histosols, Mollisols, and
Spodosols), for various reasons, play minor roles in tropical crop
production.

From the standpoint of frequency of occurrence, Alfisols are
most common, followed by Entisols and Inceptisols. Then the remaining three, Ultisols, Vertisols, and Oxisols, follow as a
closely-grouped bunch (Isbell, 1978). From the standpoint of
total area, Oxisols are most abundant, occupying 22 percent of the
tropics, followed by Aridisols (18 percent), Alfisols (16 percent), Ultisols (11 percent), Inceptisols (8 percent), Entisols (8

TABLE 8.1 (Continued on next page)
Ecoclimatic Gradient in Western and Western Equatorial Regions of Africa with Main Types of Land Use, Crops, Livestock, Grasses, and Legumes Adapted to the Regions

Approximate rainfall range (mm)	Length of dry season (months)	Zone	Vegetation	Main types of land use	Main Crops
	12	Saharan	Desert	Nomadic grazing	-
100-250	11-12	Sub-Saharan steppe	Sub-desert	Nomadic grazing	
250-600	9	Sahelian	Shrub and thorny Savannah	Semi-nomadic grazing with beginnings of semi-arid cultivation	Sorghum and millet-irrigated rice
600-1250	6-8	Sudanian	Tree Savannah	Semi-nomadic grazing with arable cultivation with fallows generally of medium to long duration	Sorghum, millet, ground nuts, yams, maize, irrigated rice
>1250	3-6	Guinean	Woodlands-deciduous forests	Arable cultivation with fallows generally of medium to long duration. Forest plantations	Sorghum, millet, ground nuts, maize, irrigated rice
>1800	Short	Guinea Equatorial	Closed rain forest	Arable cultivation with fallows generally of relatively long duration. Permanent tree crops for export and local consumption	Upland rice, plantains and bananas, yams, taro, maize, oil palm, cacao, rubber, robusta coffee

Adapted from R.O. Whyte, 1968. Grasslands of the Monsoon. Faber and Faber Ltd. London. Table 6 - Ecoclimatic gradient in western and western equatorial regions of Africa (used with permission).

TABLE 8.1--(Continued)

Approximate rainfall range (mm)	Length of dry season (months)	Zone	Livestock	Grasses +	
100-250	12	Saharan	Some camels		
	11-12	Sub-Saharan	Camels, sheep and some cattle	Cenchrus ciliaris Cenchrus setigerus	Eragrostis curvula[x]
250-600	9	Sahelian	Cattle and sheep	Andropogon gayanus Cenchrus ciliaris Cenchrus setigerus	Eragrostis curvula[x] Eragrostis superba[x] Dactyloctenium aegypticum
600-1250	6-8	Sudanian	Cattle with some sheep and goats	Andropogon gayanus Bothriochloa insculpta Brachiaria decumbens Brachiaria brizantha Brachiaria mutica Cenchrus ciliaris Panicum coloratum Setaria sphacelata[x]	Chloris gayana[x] Digitaria decumbens Digitaria smutsii Eragrostis curvula[x] Eragrostis superba[x] Panicum maximum Pennisetum purpureum Urochloa mosambicensis
>1250	3-6	Guinean	Cattle depending upon absence of tsetse, goats and some sheep	Brachiaria brizantha Brachiaria decumbens Brachiaria ruziziensis Chloris gayana[x] Panicum maximum	Pennisetum purpureum Pennisetum clandestinum[x] Setaria sphacelata[x] Melinis minutiflora
>18000	Short	Guinea Equatorial	Goats with cattle in forest-savannah mosaic or coast	Panicum maximum Pennisetum purpureum Setaria sphacelata[x]	Pennisetum clandestinum[x] Setaria sphacelata[x]

+Grasses of African origin (not necessarily West Africa) but used as sown pasture species elsewhere (R.W. Strickland, pers. comm., 1978). Those marked x are from and for higher altitudes or latitudes but have been included for completeness.

TABLE 8.1--(Continued)

Approximate rainfall range (mm)	Length of dry season (months)	Zone	Legumes++	
			African	South American
100-250	12	Saharan		
	11-12	Sub-Saharan	Lotononis bainesii^x	Stylosanthes hamata S. humilis S. scabra
250-600	9	Sahelian	Lotononis bainesii^x S. fruticosa	
600-1250	6-8	Sudanian	Glycine wightii^x Lotononis bainesii^x Macrotyloma axillare Trifolium spp.^x	Calopogonium mucunoides Desmodium intortum^x Desmodium uncinatum^x Macroptilium atropurpureum Macroptilium lathyroides	S. guianensis S. hamata S. humilis S. scabra
>1250	3-6	Guinean	Glycine wightii^x Lotononis bainesii^x Trifolium spp.^x	Capologonium mucunoides Centrosema pubescens Desmodium intortum^x Macroptilium atropurpureum	S. guianensis S. hamata
>1800	Short	Guinea Equatorial	Glycine wightii^x	Calopogonium mucunoides Desmodium intortum^x Centrosema pubescens S. guianensis	

++Legumes of South American or African origin but used as sown pasture species in other continents. Those marked x are from higher altitudes or latitudes.

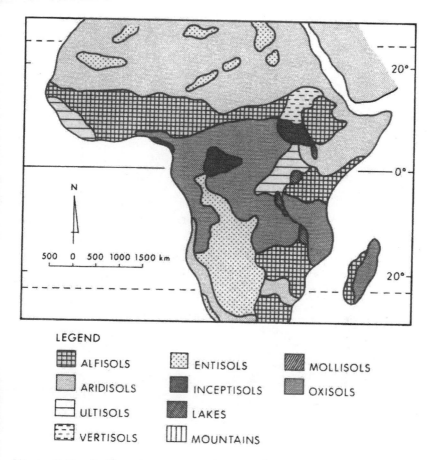

LEGEND

	ALFISOLS		ENTISOLS		MOLLISOLS
	ARIDISOLS		INCEPTISOLS		OXISOLS
	ULTISOLS		LAKES		
	VERTISOLS		MOUNTAINS		

Figure 8.2. Soil orders of tropical Africa.

percent), and Vertisols (2 percent).[1]

Highly Weathered and Leached Soils

Many of the malnutrition problems in dairy and beef animals in the tropics are associated with low quality pasture grasses growing on poor quality lands. Poor quality lands are almost synonomous with highly weathered and leached acid-infertile soils. These soils are predominantly Oxisols and Ultisols in tropical

---

[1]Calculated by M. Drosdoff, Cornell University, on the basis of Aubert and Tavernier's (1972) map. Data in Sanchez (1976).

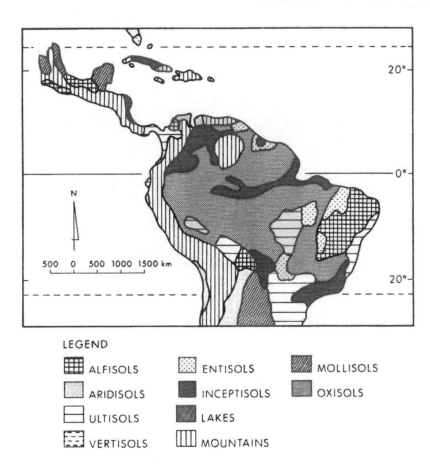

LEGEND

⊞ ALFISOLS		▦ ENTISOLS		▨ MOLLISOLS	
▨ ARIDISOLS		■ INCEPTISOLS		▨ OXISOLS	
⊟ ULTISOLS		▨ LAKES			
▨ VERTISOLS		⦀ MOUNTAINS			

Figure 8.3. Soil orders of tropical Latin America.

America, Southeast Asia, and Africa and, to a lesser extent, Alfisols in Africa and India.

    Oxisols. Oxisols are usually found on slightly sloping to
level, stable landscapes in the rainy, humid-seasonal, and wet-dry
tropics (4.5 to 12 months with more than 100 mm rainfall). Native
vegetation ranges from high rain forest and anthropic savanna with
vegetation which drops its leaves during the dry season, to savanna and grassland with deciduous woody plants in scattered groves.
Most Oxisols are deep, friable, well-drained but infertile red to
yellow soils composed of colloidal oxides and hydroxides of aluminum and of iron and other heavy metals, plus quartz. In texture,

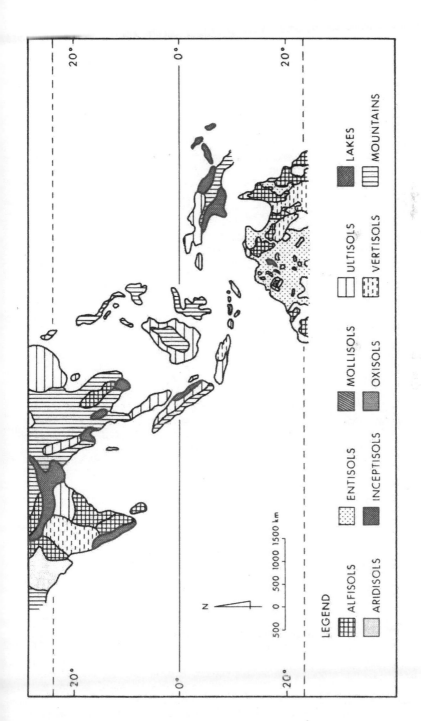

Figure 8.4. Soil orders of tropical Asia.

most Oxisols are non-sticky, non-expanding clays. These soils do not crack upon drying, are usually well-drained and have the water infiltration capacities of sandy soils. Oxisols are usually found in association with and share many of the characteristics of the Ultisols.

Ultisols. These soils often are found in the same landscape position as are Oxisols, though they may be found on steeper slopes of areas adjacent to those occupied by Oxisols, i.e. valley edges. The major difference between Ultisols and Oxisols is that the former have a horizon enriched with silicate clays within a short distance of the surface. This implies that digging to no more than a meter in depth should uncover a horizon in an Ultisol containing structural soil components with noticeable clay skins or slickensides. Since the clay composition is usually of the kaolinitic non-expanding type plus oxides and hydroxides of iron and aluminum as found in the Oxisols, cultivation of the Ultisols should not be more difficult than that required for the Oxisols.

Alfisols. Oxisols and Ultisols are often found in association with Alfisols with the latter also occupying stable but more steeply sloping areas than do the Oxisols. Alfisols are more fertile than Ultisols, and, in color, grade from red and yellow into brown and grey. They can be managed much as the Ultisols except when there is a layer of indurated laterite (plinthite) present which could cause the soil above this layer to be much more easily eroded than is the case for Alfisols, Ultisols, or Oxisols lacking plinthite.
Together, the highly weathered and leached soil (Oxisols, Ultisols, and Alfisols) of the tropics cover an area of 25 million square kilometers. This is an area only slightly smaller than the total land area of Europe plus North America.
It should be noted that the identification of two soil horizons, argillic and oxic horizons, serves as an important diagnostic tool in determining in what soil order a particular soil should be placed. The identification of such horizons is a major problem in many highly weathered tropical soils (Isbell, 1978).

Continental Distribution of Soils

Six major regions, as delineated by Isbell, 1978, are used to show the continental distribution of major soil orders in the tropics. They are:

South America. Oxisols and Ultisols occur extensively in South America, including the Cerrado, the Llanos and the eastern Amazon basin. They occur extensively under humid conditions (greater than 1000 mm annual rainfall), although extensive areas of Oxisols are also found in northeast Brazil under subhumid conditions (500-1000 mm rainfall). Alfisols appear to be dominant in subhumid regions, such as in northeast Brazil, the north coast of Colombia, and western Venezuela. Entisols are found in close association with coarse-textured Oxisols in humid areas, such as in

central and eastern Brazil, and also found extensively as shallow
soils throughout the steeper parts of the Andean, as well as other
mountainous areas. Most of the Inceptisols are concentrated as
volcanic ash soils in the Andes.

Central America-Caribbean. Inceptisols occur extensively in
mountainous areas of Mexico, Central America, and the Caribbean
and are associated with the presence of volcanic ash. Alfisols
occur commonly in subhumid areas together with Vertisols and Mollisols, with the latter of local importance, such as in the Yucatan Peninsula.

Africa. Oxisols and Ultisols are found extensively in humid
areas and are associated with old stable surfaces. Alfisols are
also found in the humid zone but in the lower rainfall limits.
They are most common in the subhumid region. Also widespread in
the subhumid region is the Entisol, e.g. in the Sahelian Zone and
Angola. Vertisols are of common occurrence in subhumid areas,
too. A prime example would be in Sudan on alluvial material.
Some Inceptisols are associated with youthful unstable surfaces.
Aridisols are most extensive in arid areas.

Indian Subcontinent. Alfisols occur widely throughout India
in both humid (but toward lower rainfall limits) and subhumid
climates. Vertisols are also found in abundance in subhumid climates. Both of these soil orders are associated with basic parent
materials. Some Inceptisols are widespread in places such as the
Ganges alluvial plain. Oxisols and Ultisols are less prominent in
India than in South America and Africa.

Southeast Asia-Indonesia-New Guinea. Ultisols dominate upland areas of Southeast Asia. Inceptisols appear extensively in
younger geomorphic areas. Vertisols dominate much of eastern
Java. Alfisols are found in upland areas of Thailand. Histosols
are of local importance in certain low-lying poorly-drained areas
and are often associated with acid-sulfate clays. There is no
widespread occurrence of Oxisols.

Australia. Entisols occur extensively in humid and subhumid
regions of tropical Australia, occurring mainly as shallow soils,
usually derived from siliceous sandstone. Alfisols dominate the
subhumid regions, together with Vertisols. Oxisols and Ultisols
are of less importance than in South America or Africa. Aridisols, as in Africa, occur extensively in arid areas.

Soils Properties

Physical. The water-holding capacity of a tropical soil assumes a very important role, especially in the subhumid and arid
areas of the tropics, where drought periods are frequent and
lengthy. Aridisols and Entisols generally exhibit a low water
storage capacity because they are usually shallow and sandy. The
fact that a typical soil is low in sand and high in clay does not,

however, automatically assure a much improved water storage capacity. For example, Oxisols are generally characterized by strong granulation, which means improved physical structure for tillage, but also means that they behave like a sand in water retention. Some oxic subgroups of Ultisols and Alfisols also show this "sandy" characteristic. Fortunately, many of these latter soils, as well as Inceptisols, are deeply weathered and this fortuitous circumstance means improved water storage. Also, because of the strong aggregating nature of these soils, permeability and drainage features are good to excellent. Pan formation which restricts water movement and root penetration does pose a problem in some of these soils, especially in Oxisols. But this problem is by no means as great as earlier thought to be when laterite formation was erroneously believed to be dominant in the tropics.

In contrast to the foregoing soils, the Vertisols behave truly as heavy clays. These soils can hold a large amount of water; however, the percent of moisture at wilting point is also quite high, which means that the range of available moisture between water-holding capacity and wilting point is narrow. These soils are also difficult to cultivate because of their shrink-swell properties. These soils are relatively impermeable and internal drainage is poor. The combination of such poor features, coupled with location of these soils generally in low-lying, low-rainfall areas, often produces salinity problems.

Chemical/Fertility. Oxisols and Ultisols are the most acid soils of the tropics, together with some Inceptisols and Histosols. Soil pH values can drop below 4.0 in extreme cases, especially in areas under cultivation, and fertilization. Most Alfisols and Entisols are moderately acid, while Vertisols and Aridisols are near-neutral to alkaline.

Associated with very acid soils are problems of high aluminum concentrations, some to the extent of adversely affecting plant growth. Excessive manganese can also be present, although this problem is usually restricted to Oxisols. Calcium and magnesium deficiencies, when found, are generally also associated with acid, highly weathered soils.

Oxisols, Ultisols, and some Inceptisols are regarded as high fixers of phosphorus, mainly because of the presence of high amounts of iron and aluminum oxides in many of these soils; thus, phosphorus deficiency is most likely to be found in these three soils. Alfisols and Entisols are generally associated with moderate phosphorus deficiency. Sulfur deficiency is often found along with phosphorus deficiency and can be widespread in Alfisols and Entisols.

Potassium deficiency is most likely to be related to the mineralogical make-up of a tropical soil. Soils that are derived from kaolin and iron/aluminum oxides are associated with potassium deficiency, whereas those from micaceous and related minerals will not require potassium fertilization.

Micronutrient deficiencies (molybdenum, zinc, copper, and boron) have been investigated on a world wide network of benchmark soils in the tropics (Swindale, 1978).

Soil-Tropical Legume Relationships

A great potential exists to increase milk and beef production through the use of improved pasture legumes grown on acid-infertile tropical soils. Russell (1978) has summarized two approaches that can be taken to improve growth in these soils. The first is to improve soil conditions by applying fertilizers or lime. The second is to select or breed or both, legumes capable of withstanding limiting soil conditions. Most situations call for both approaches.

To investigate soil management practices for improved legume growth in the first approach, it is necessary that we understand the soil mineralogy of these soils. Uehara (1978) stated that many of the acid-infertile soils of the tropics are made up of low activity clays which are pH-dependent charge minerals and belong to the constant charge surface potential group. Where tropical pasture legumes are grown in soils dominated by low activity clays, as in Oxisols and Ultisols, lime is added to supply nutrients and not as an amendment (as contrasted to high activity clay soils where lime is added to correct aluminum toxicity and to raise soil pH above a critical value). Correcting phosphorus deficiency in low activity clay soils can become costly because these clays fix relatively high amounts of soluble phosphorus fertilizers. Similarly, if these soils are deficient in heavy elements, like zinc or copper, treatments will call for rather large rates to overcome specific adsorption of ions. An important redeeming feature in these clays is the formulation of stable aggregates that endows the soils with good physical properties.

In the second approach we must proceed on the basis that tropical pasture legumes to be used effectively on acid-infertile soils must be selected or bred for their adaptability to grow under conditions of high soluble aluminum (and likely manganese, too), low exchangeable calcium and magnesium, low available phosphorus, and drought and flood stress. In addition to these soil condition requirements, the legume should effectively nodulate with Rhizobium, be compatible with grasses, tolerate disease and insect attacks, and be palatable.

Important tropical legumes that appear to tolerate acid-infertile soils are of the genera: Stylosanthes, Desmodium, Centrosema, Pueraria, and Macroptilium (Kerridge, 1978). Zornia, Aeschynomene, and Galactia are added to this list by Schultz-Kraft and Giacometti (1979). They also include Calopogonium, Rhynchosia, Cassia, Mimosa, and Tephrosia as genera deserving of future attention for adaptation in these soils. Adaptive pattern differs considerably between species within these genera and between the above-mentioned genera and other tropical genera (Kerridge, 1978).

The genus Stylosanthes has received the greatest attention because of its adaptability to acid-infertile soils (Kerridge, 1978; Schultze-Kraft and Giacometti, 1979). This genus appears the best among tropical pasture legumes in tolerating high aluminum and manganese, and low phosphorus and molybdenum. Stylosanthes species that adapt well to Oxisols and Ultisols of South

America are: S. guianensis, S. scabra, S. viscosa, S. humilis, and S. capitata. The latter deserves special mention because it is the only Stylosanthes species that appears to sufficiently tolerate anthracnose and stem borer attacks (Schultze-Kraft and Giacometti, 1979).

Among Desmodium, D. barbatum and D. cajanifolium are the two species most frequently found growing in regions of acid-infertile soils. Of species of Desmodium native to southeast Asia which have been tested on Oxisols and Ultisols of tropical America, Desmodium ovalifolium appears to show best tolerance to acid-infertility and drought (Schultze-Kraft and Giacometti, 1979).

Although Centrosema has a wide geographic distribution, most Centrosema species are native to regions with relatively high soil fertility. There are notable exceptions (see Section II) and with further collections the genus may have a much wider range of adaptation than previously thought.

To obtain optimum growth in these acid-infertile soils, it is necessary that an effective legume-Rhizobium symbiosis be obtained. Halliday (1979) reported that most tropical pasture legumes do not nodulate freely in acid soils. Hence strains of Rhizobium must be selected that are compatible with the host plants adapted to these soils.

EFFECTS OF CLIMATE AND SOIL ON PLANT GROWTH AND ANIMAL PRODUCTION

In general, the tropics are regions of high temperature. When soil moisture content is adequate, plant growth rates are high. This is particularly true of the grasses, many of which have a special pathway of photosynthesis which allows them to rapidly respond to such conditions. This rapid growth, particularly when linked with low soil-mineral levels, "dilutes" the mineral content of the plants which then tend to be low in phosphorus, sulfur, etc. Animals can sometimes compensate for this by careful selection of their diet. Unfortunately, this compensation may not be sufficient therefore mineral deficiencies in animals can occur. During the wet season large animal weight gains can be recorded.

In the dry season, however, plant growth ceases, protein levels decline and much of the protein and minerals accumulated by the plants are translocated to the root systems. Animals live on standing hay, often of low feeding value, and the quality drops with time. Out-of-season rains are often insufficient to promote renewed growth and may reduce feed quality even further. Animals lose weight, often have their potential fertility level reduced, and sometimes even die. Thus the potential of tropical grazing lands is often limited by the carrying capacity of the land in the dry season, although in the wet season it may be undergrazed.

There are numerous ways of tackling this feed problem. Basically there are three alternatives: 1) remove the excess feed in the growing season and either store it for use in the dry season or use it for some other purpose; 2) remove some or all of the animals in the dry season; 3) make more feed available in the dry season which can be done by the feeding of hay or supplements.

A long-lasting solution to the shortage in the dry season may lie in the provision of better pasture species, grasses which grow longer into the dry season and respond quickly to the opening rains of the season, and legumes which stay green and provide a high protein diet which allows the animals to utilize the poor quality standing hay for energy.

REFERENCES

Aubert, G. and Tavernier, R. 1972. Soil Survey. In "Soils of the Humid Tropics" National Academy of Sciences, Washington, D.C.

Halliday, J. 1979. Field responses by tropical forage legumes to inoculation with Rhizobium. In "Pasture Production in Acid Soils of the Tropics" (eds.) P.A. Sanchez and L.E. Tergas. CIAT, Cali.

Hutton, E.M. 1979. Problems and successes of legume-grass pastures, especially in tropical Latin America. In "Pasture Production in Acid Soils of the Tropics" (eds.) P.A. Sanchez and L.E. Tergas. CIAT, Cali.

Isbell, R.F. 1978. Soils of the Tropics and Sub-Tropics: Genesis and Characteristics. In "Mineral Nutrition of Legumes in Tropical and Subtropical Soils" (eds.) C.S. Andrew and E.J. Kamprath. CSIRO, Melbourne.

Kerridge, P.C. 1978. Fertilization of acid tropical soils in relation to pasture legumes. In "Mineral Nutrition of Legumes in Tropical and Subtropical Soils" (eds.) C.S. Andrew and E.J. Kamprath. CSIRO, Melbourne.

Landsberg, H.E., Lippman, H., Paffen, K.H., and Troll, C. 1966. In "World Maps of Climatology" (eds.) E. Rodenweldt and H.J. Jusatz, Springer-Verlag, Berlin.

Russell, J.S. 1978. Soil factors affecting the growth of legumes on low fertility soils in the tropics and sub-tropics. In "Mineral Nutrition of Legumes in Tropical and Subtropical Soils" (eds.) C.S. Andrew and E.J. Kamprath. CSIRO, Melbourne.

Sanchez, P.A. 1976. Properties and Management of Soils in the Tropics. John Wiley and Sons. New York.

Sanchez, P.A. and Isbell, R.F. 1979. A Comparison of the Soils of Tropical Latin American and Tropical Australia. In "Pasture Production in Acid Soils of the Tropics" (eds.) P.A. Sanchez and L.E. Tergas. CIAT, Cali.

Schultze-Kraft, R. and Giacometti, D.C. 1979. Genetic resources of forage legumes for the acid, infertile savannah of tropical America. In "Pasture Production in Acid Soils of the Tropics" (eds.) P.A. Sanchez and L.E. Tergas. CIAT, Cali.

Spain, J.M. 1979. Pasture establishment and management in the Llanos Orientales. In "Pasture Production in Acid Soils of the Tropics" (eds.) P.A. Sanchez and L.E. Tergas. CIAT, Cali.

Swindale, L.D. 1978. A Soil Research Network Through Tropical Soil Families. In "Soil Resource Data for Agricultural Development" (ed.) L.D. Swindale. Univ. of Hawaii, Honolulu.

Uehara, G. 1978. Mineralogy of the predominant soils in tropical and sub-tropical regions. In "Mineral Nutrition of Legumes in Tropical and Subtropical Soils" (eds.) C.S. Andrew and E.J. Kamprath. CSIRO, Melbourne.

USDA, SCS, Soil Survey Staff. 1975. Soil Taxonomy: A Basic System of Soil Classification for Making and Interpreting Soil Surveys. Agric. Handbook No. 436. Washington, D.C.

Whyte, R.O. 1968. Grasslands of the Monsoon. Faber and Faber, Ltd. London, England.

Section II
Characteristics of *Centrosema,*
Desmodium, and *Stylosanthes*

1
Introduction

P. P. Rotar

Prior to the middle 1950's, there was very little work directed toward tropical pasture improvement; many early reports
were about observations from introduction gardens. These reports
were often vague and said little more than that the accession(s)
looked promising for further development. Moreover, there were
only a handful of research stations directly involved in tropical
and subtropical pasture improvement. Fewer still had active tropical and subtropical forage breeding programs. The efforts put
into tropical pasture legume research may be measured in part by
the increase in the number of publications that have appeared in
the world's literature. Extensive manual and computer based
searches for the agronomic literature pertaining to Centrosema,
Desmodium, and Stylosanthes have been made for the period 1955-
1976.[1] For the most part, taxonomic references found in the mul-
titudinous flora's published throughout the world are excluded.
These results have been tabulated and are summarized in Figures
1.1-1.5. The numbers presented are approximate, as they omit information published locally that is only rarely included in world
data bases. Slightly more than 1500 citations have been retrieved
according to the date of publication for the three genera - Centrosema, Desmodium, and Stylosanthes and their species during the
21-year period 1955-1976. Nearly 49 percent of the references
pertained to the genus Stylosanthes, 19 percent to the genus Centrosema, and 32 percent to the genus Desmodium. These percentage
data are summarized graphically in Figure 1.
  Figure 1.2 shows the increase in the number of publications
pertaining to the three genera during the period 1956-76. The
increasing numbers of citations in the literature for these three
genera reflect not only the interest in identifying tropical forage legumes but also are indicative of the information explosion
confronting agricultural scientists worldwide. Very few citations
were found prior to 1966, but a substantial yearly increase occurred during the period 1966-1976.

---

[1] These citations are being published in bibliographic form and are
available in microfiche (Rotar, et al., 1981).

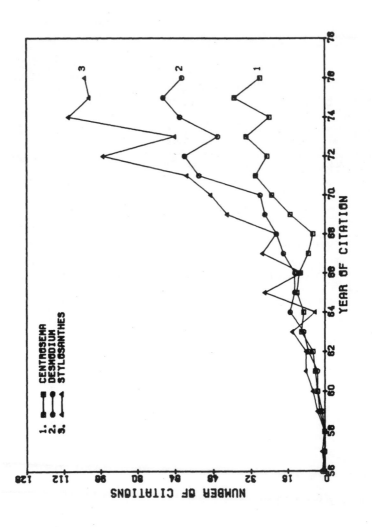

Figure 1.1. The number of literature citations, world-wide, for the genera _Centrosema_, _Desmodium_, and _Stylosanthes_ during the period 1956-76.

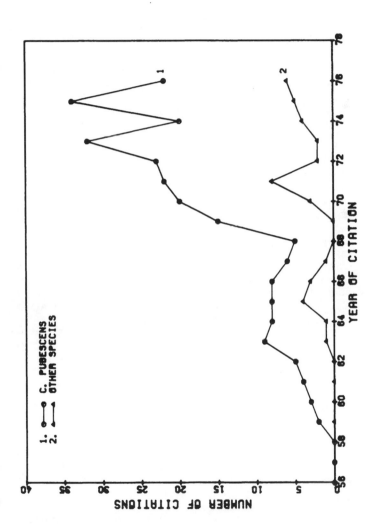

Figure 1.2. The number of literature citations, world-wide, for the genus _Centrosema_ during the period 1956-76.

Figures 1.2, 1.3 and 1.4 present the trend in number of literature citations for the species within each of the three genera. The number of publications containing information on Centrosema species (Figure 1.2) began increasing in 1962, but the marked increase commenced in 1968. During the entire period surveyed, however, one species - Centrosema pubescens received more attention than all other Centrosema species combined. In Figure 1.3, we see that three Desmodium species (D. intortum, D. sandwicense, and D. uncinatum) are cited more often than the rest. In Figure 1.4, we see a similar situation, three of the Stylosanthes species (S. guianensis, S. hamata, and S. humilis) were referenced more often than the remainder of the genus. Work on S. humilis and S. guianensis has been carried on since before 1960 and intensively since 1966. Considering that the genus Centrosema contains perhaps as many as 50 species, Desmodium, 350 species, and Stylosanthes, 70 species, only a portion of the germplasm of these genera has been evaluated agronomically. Of that receiving attention, the figures presented suggest that emphasis has flunctuated between research and publication, and possibly in the amounts of funds made available for research activities.

Figure 1.5 clearly shows that a good percentage of the citations pertaining to the three genera are of Australian origin. It also shows the developing interest of CSIRO scientists in tropical and subtropical pasture development beginning with the establishment of the Division of Tropical Pastures - the Cunningham Laboratory at Brisbane which has expanded to the Division of Tropical Crops and Pastures with an additional pastoral laboratory; The Davies Laboratory at Townsville, Queensland. We should point out that the various Australian Commonwealth Departments of Primary Industries also contributed substantially to these efforts.

Currently, there are several International Agricultural Research Centers and many National Agricultural Research Centers devoting considerable time and effort to the development of tropical pastures. Much of the impetus for these activities was based upon the Australian successes. A less desirable consequence of the increased awareness of potential benefits from the agronomic application of research results is the difficulty in keeping current with information that is becoming available.

Descriptive titles and adequate key word listings for all research are basic requirements for information retrieval. Scientific names should be presented for inclusion in documentation identifying the article's contents. Common names are seldom of value by themselves in computerized searches for information retrieval; however, where possible, such information should be included along with scientific species identification, because such knowledge may be useful to those communicating information derived from research to the local user who may be a herdsman or grazier.

Section II will present reviews of the three genera Centrosema, Desmodium, and Stylosanthes as they have been investigated for use in pastures. The reviews include information about their taxonomy, species of agronomic value, breeding behavior, and genetics where known and their adaptation. The reviews are followed by a chapter on Rhizobium requirements for the genera.

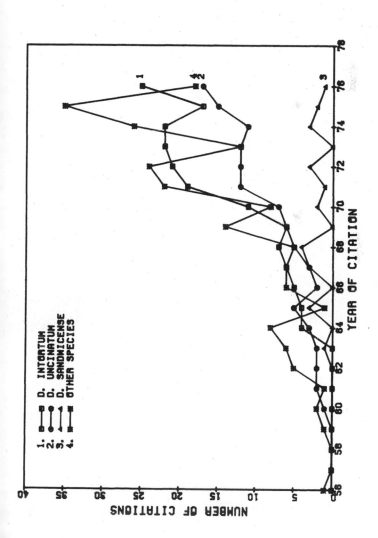

Figure 1.3. The number of literature citations, world-wide, for the genus _Desmodium_ during the period 1956-76.

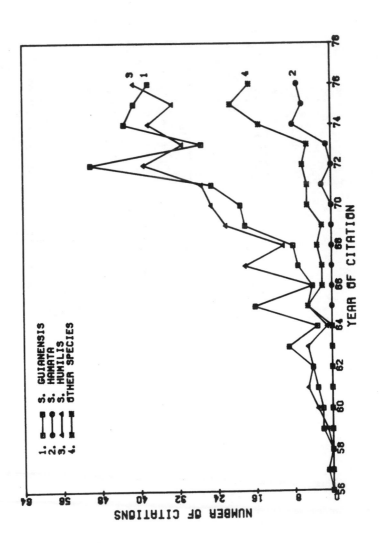

Figure 1.4. The number of citations, world-wide, for the genus _Stylosanthes_ during the period 1956-76.

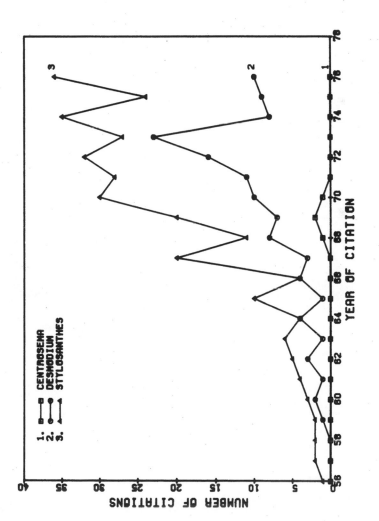

Figure 1.5. The increase in number of literature citations from Australia for the three genera <u>Centro-sema</u>, <u>Desmodium</u>, and <u>Stylosanthes</u> during the period 1956-76.

REFERENCES

Rotar, P.P., Harris, S., Evans, D., and Walker, J.L. 1982. Se-
    lected bibliography of Centrosema, Desmodium, Stylosanthes
    and other tropical and subtropical pasture legumes. Univer-
    sity of Hawaii, Hawaii Institute of Tropical Agriculture and
    Human Resources Research Series (in press).

2
Centrosema

R. J. Clements, R. J. Williams,
B. Grof, and J. B. Hacker

INTRODUCTION

Centrosema pubescens has a longer history of domestication than most tropical forage legumes. It was introduced to Southeast Asia from tropical America during the 19th century or earlier, and in the early 20th century became an important cover crop in rubber plantations. It was also one of the first legumes to be developed for sown tropical pastures, following Schofield's (1941) pioneering research on the wet tropical coast of Queensland, Australia. Consequently, for the cultivated form of this species there is a considerable amount of published literature (at least 300 references) relating to its broad climatic/edaphic adaptation, agronomy and management, and ability to fix atmospheric nitrogen in suitable habitats. Although there has been no comprehensive review of the literature, there are numerous popular descriptions of C. pubescens and its cultivation, and its essential features are well known.

On the other hand, knowledge of the other Centrosema species has been fragmentary until the present time. Information is now being obtained rapidly but most of it is still unpublished and incomplete, and many of the present ideas may eventually need some revision. However, the experience gained from studying the full range of species has altered our perspectives and expectations of the genus as a whole.

In this overview special emphasis will be placed on the lesser-known species, particularly those showing economic potential. When considering C. pubescens, attention will focus on aspects of the productivity and nitrogen fixing ability of this legume and on factors which limit its economic capabilities.

SPECIES DIVERSITY IN CENTROSEMA

Centrosema is native to Central and South America, the Caribbean area, and southern U.S.A., and several species have become naturalized in tropical Africa, Southeast Asia, and India. The genus is closely related to Periandra, and to Clitoria, from which it was separated in 1837 by Bentham. All three genera are members of the tribe Phaseoleae, which also includes Glycine, Phaseolus, Vigna, and Dolichos. Specimens of Centrosema have occasionally been misclassified into these related genera.

More than 100 species names have been referred to Centrosema, but only about 35 species are now recognized (Clements and Williams, 1980). There is no monographic treatment of the whole genus, although a complete revision is planned (Williams and Clements, in preparation). There are several regional treatments, but with the notable exception of a recent description of the Brazilian species (Barbosa-Fevereiro, 1977) they are so outdated that they are actually misleading. Until a complete revision is published, Barbosa's names, (which include about half of the total number of species), should be used.

Most Centrosema species are tropical, viny perennials, but the genus extends into many subtropical and some maritime climates as defined by Papadakis (1966), and includes some subshrubs and annuals. The range of native habitats includes forest fringes in the high-rainfall tropics; humid and sub-humid savannas and woodlands; deciduous semi-arid to arid forests and scrub; coastal sand dunelands; and subtropical pine forests. Some species extend more or less continuously for hundreds of kilometers, while others are known from a single location or are restricted to a number of disjunct locations. The two main centers of species diversity are the Brazilian Central Plateau and Central America (Clements and Williams, 1980); these regions account for about three-quarters of the reported species.

It is difficult to make broad generalizations about the adaptation of Centrosema, because many of the species are adapted to contrasting environments, and because the species that have been well studied each contain diverse forms. A few examples will illustrate this point. C. brasilianum is most common in hot tropical and semi-arid tropical environments (as defined by Papadakis, 1966), yet it has been collected from the margins of the Amazon rainforest where the annual rainfall exceeds 2000 mm. C. pascuorum is restricted to a much narrower climatic range, but occurs on a very wide range of soils. C. virginianum is so variable that some forms can tolerate a temporary snow cover while other survive in the semi-arid tropics of north-eastern Brazil. One of the few generalizations that can be made is that Centrosema is not simply a "wet tropics" genus. This false reputation is based on agronomic experience with the genetically restricted, cultivated form of C. pubescens. Although it is true that C. pubescens is primarily adapted to the wet tropics, (e.g. humid, semi-tropical climates of Papadakis, 1966), it has been collected from 500-600 mm rainfall regions on the southern margin of the Guajira Peninsula of Colombia and from areas above 1000 m altitude near Santa Cruz, Bolivia (18°S).

A list of the species having commercial value or at least some agronomic potential requires some judgment, because our information on many species is extremely limited. It is also based largely on Australian experience, although field trials in the U.S.A. (Kretschmer, 1977) and unpublished reports from Colombia,

TABLE 2.1
Chromosome Numbers of Centrosema Species

SPECIES	NO. OF ACCESSIONS COUNTED	ORIGINS OF ACCESSIONS[1]	CHROMOSOME NUMBER (2n)
C. pubescens	16	Br, M, V, Gt, Gn	22
C. virginianum	12	Br, M, V, H, Bo, P	18
C. brasilianum	4	Br	22
C. plumieri	2	D, F	22
C. sagittatum	3	Br, A	22
C. schottii	3	M, Co	22
C. pascuorum	6	Br, E	22
C. angustifolium	1	Br	22
C. grandiflorum	2	Br	22
C. sp. aff. acutifolium	3	Br	22
C. sp. aff. pubescens[2]	1	Cr	22
C. macrocarpum	1	Co	22

[1] Br (Brazil); M (Mexico); V (Venezuela); Gt (Guatemala); Gn (Guyana); H (Honduras); Bo (Bolivia); P (Paraguay); E (Ecuador); A (Argentina); Co (Colombia); Cr (Costa Rica); D (Dominican Republic); F (Fiji-original source unknown).
[2] cv. Belalto

Malaysia, and Thailand provide valuable confirmation in some cases. Such a list must obviously include the two commercially developed species C. pubescens, C. plumieri, and also C. pascuorum. C. virginianum may eventually become a commercially useful plant in the subhumid subtropics, but some breeding work is needed to combine the desirable features of the tropical and subtropical forms. C. brasilianum and C. schottii have shown promise in Australia. The group of species closely related to C. pubescens (Clements and Williams, 1980), and also the three species C. sagittatum, C. arenarium, and C. rotundifolium have not been extensively collected by agricultural plant explorers and deserve greater attention. In particular, the potential of the latter two species in semi-arid tropical environments should be evaluated as soon as suitable collections are available. In summary, about one-third of the species appear to have some agricultural significance.

Most of the species for which living material has been examined appear to be self-pollinated (cleistogamous) though some open-pollination (out-crossing) does occur (eg. in C. macrocarpum and C. arenarium; R. Schultze-Krafte, pers. comm.). Some interspecific hybrids have been reported; however, some of these are in fact intraspecific hybrids, usually resulting from crosses within C. pubescens. J.B. Hacker (Table 2.1) has established that many species (including C. pubescens, and C. pascuorum) have 22 somatic chromosomes while C. virginianum has 18. The only previously reported chromosome count (C. pubescens, 2n = 20; Frahm-Leliveld, 1953 appears to be incorrect.

GENE BANKS

The first point for emphasis is that current gene banks (germplasm collections) are grossly inadequate. Many species are completely unrepresented in existing collections and several significant geographical areas have not yet received sufficient attention from plant collectors.

The major collections are increasing quite rapidly and estimates of numbers of specimens (accessions) are soon outdated; however, some idea of the size of these collections may be obtained from Table 2.2. Because there is considerable overlap and sharing of collections, the total number of unique accessions is much less than the sum of all collections and is probably little more than 700 at the present time (May, 1979). Storage conditions vary considerably but are generally adequate to maintain seed viability for several years. In Australia, maintenance and regeneration of seed stocks is given high priority in plant introduction and breeding programs. The recent rapid rate of growth of the total Australian collection (CSIRO plus Queensland Department of Primary Industries) is apparent from the following counts:

	Accessions	Species
Pre 1960	32	3
December, 1965	98	4
December, 1967	173	10
December, 1971	277	10
August, 1976	312	13
January, 1977	328	13
January, 1978	456	14
January, 1979	557	16

The species representation of the various collections is uncertain because many accessions have not been identified, partly due to the inadequacies of the currently available taxonomy; however, the structure of the Australian collection (Table 2.3) is probably fairly typical. It will be seen that C. pubescens and C. virginianum comprise the bulk of the collection. From herbarium studies and evaluation of early living material, we know that the C. virginianum collection, though small, is nearly representative of the genetic diversity within this species. However, the extreme narrow-leafed (angustifoliate) form which occurs in the southern U.S.A. and parts of the Caribbean region is not represented in the collection.

The C. pubescens accessions are more numerous but are less representative. Many of them have been obtained by correspondence, and some of these are simply re-introductions of the narrowly based commercial material. Although about three-quarters of the C. pubescens accessions are native to their collection site, some morphological forms and known areas of occurrence are poorly represented in the Australian collection.

The accessions of the other species listed in Table 2.3 are not at all representative of the intraspecific diversity. The small Australian collection of C. pascuorum represents most of its known areas of occurrence, and contains forms not represented in old-world herbaria (Williams and Clements, unpublished data).

Major factors to consider when setting plant collection priorities are: current status of these collections, list of promising species, distribution patterns of these species, experience with accessions from various locations, and interests and geographical locations of scientific organizations. It must be stressed that the total area of Centrosema distribution is so large and so inadequately explored that no single group of scientists can hope to cover more than a fraction of it. At the present time, both agricultural scientists and plant taxonomists visit the various regions within the total area. There is a great need for a systematic and mutually helpful approach to this collecting activity. Cooperation between taxonomists and agriculturalists is highly desirable; each scientist should collect material suitable for the other.

There is a risk that unless adequate feedback and cooperation occurs between groups of collectors, the benefit of any increase in collecting activity will be temporary. Duplication of effort will result, and valuable collections will be lost as short-term projects finish and local evaluation is completed. Many collections have already been lost. A thorough understanding of the taxonomy, distribution, adaptation, and species relationships of the genus will only be obtained efficiently when a few major gene centers gain access to all collections. Only a network of major centers collaborating with local evaluation groups can ensure maintenance of all the material.

Bearing in mind the diverse interests of agricultural scientists, taxonomists, and others, it is believed that collecting within the following geographic areas should receive priority:

1. Central America. Attention should be concentrated on C. pubescens (and its close relatives) and C. pascuorum. CSIRO plans further collection in this area in the near future.

2. Brazilian Central Plateau (particularly the States of Minas Gerais, Mato Grosso, Goias and Bahia). This area contains great species diversity. Collections would be particularly useful for taxonomic studies and would enable some agronomic evaluation of a number of species which have not so far been studied (e.g. C. arenarium, C. rotundifolium). Diverse forms of C. pubescens and related species occur in this region.

3. Northern Bahia and north-eastern Brazil. This area has proved to be a very good gene source for species adapted to semi-arid conditions, particularly C. pascuorum, C. brasilianum, and C. rotundifolium.

4. Eastern Andean Slopes (Colombia, Ecuador, Peru and Bolivia). In this poorly explored area at 600-2000 m altitude there are interesting ecotypes of C. pubescens. The region also contains some of the taxonomically outlying species of Centrosema and is therefore significant for studies on taxonomy and patterns of adaptation.

TABLE 2.2
Number of <u>Centrosema</u> Accessions Held in Major Gene Banks

Institute	No. of Accessions	Most Recent Information
CSIRO Division of Tropical Crops and Pastures, Australia	557	March 1979
CIAT (Centro Internacional de Agricultura Tropical), Colombia	335	March 1979
Fort Pierce Agricultural Research Center (University of Florida), U.S.A.	212	February 1979
Instituto de Pesquisas IRI, Matao, Sao Paulo, Brazil	207	November 1977
EMBRAPA (CENARGEN and associated centers), Brazil	155	March 1979
Queensland Department of Primary Industries, Brisbane Australia	49	February 1979

5. <u>The Chaco Region</u> (Bolivia, Paraguay, and northern Argentina). Unusual forms of <u>C. virginianum</u>, <u>C. brasilianum</u>, <u>C. vexillatum</u>, and <u>C. angustifolium</u> occur here.

Apart from these major areas, a number of more restricted locations deserve investigation. These include inland Guyana (<u>C. pascuorum</u> and <u>C. brasilianum</u>), the Guajira Peninsula area of Colombia and Venezuela (<u>C. pubescens</u>); coastal Ecuador (<u>C. schottii</u> and <u>C. pascuorum</u>); and the southern States of Brazil (<u>C. grandiflorum</u>). The significance of <u>C. grandiflorum</u> lies in its close relationship to the more tropical <u>C. pubescens</u>, and the possibility of obtaining agriculturally useful interspecific hybrids.

TABLE 2.3
Number of Accessions of Individual Centrosema Species Held by
CSIRO and QDPI in March, 1979

SPECIES	NO. OF ACCESSIONS
C. pubescens	269
C. virginianum	92
C. plumieri	58
C. schottii	35
C. brasilianum	17
C. pascuorum	12
C. rotundifolium	9
C. angustifolium	6
C. sp. aff. acutifolium	6
C. sagittatum	5
C. macrocarpum	4
C. schiedeanum	2
C. grandiflorum	2
C. arenarium	1
C. bracteosum	1
C. venosum	1
Unidentified accessions[1]	37
TOTAL	557

[1]Mainly old accessions (probably C. pubescens) and recent accessions not yet quarantined.

C. PUBESCENS

In this section comments are restricted to the "common" or commercial forms of C. pubescens, known as centro. As already mentioned, these forms give no idea at all of the diversity of the species, and any conclusions about the adaptation of the commercial forms do not necessarily apply to the species as a whole.

The cultivation of centro as a cover crop in plantation agriculture has been described in numerous articles (e.g. Anon. 1954, 1958). In particular a great deal of information was summarized by Watson and colleagues in the "Planters' Bulletin" no. 68 (1963). The cultivation of centro as a pasture plant has also been well outlined. Teitzel and Burt (1976) have recently summarized Australian experience with this legume, and there are several other more general descriptions of its adaptation, establishment, and husbandry (Ahlgren, 1959; Toutain, 1973; Humphreys, 1974; Bogdan, 1977; Skerman, 1977). The two most recent references in particular provide a useful popular guide to the use of centro in pastures.

Centro is widely cultivated as a ground cover in rubber,

Figure 2.1. C. pubescens with flowers and seed pods.

coconut, and oil palm plantations in Southeast Asia, India, the
Pacific region, and to a limited extent in Africa. The contribution of centro to these cropping systems is difficult to quantify,
but plantation crops are of great economic importance in the tropics and occupy extensive areas. For example, there are about 5.5
million hectares of rubber planted in Southeast Asia alone. A
mixture of centro, puero (Pueraria phaseoloides), and calopo (Calopogonium mucunoides) is used routinely as a cover crop in rubber
plantations, where it persists for 3-6 years until the tree canopy
is complete. Since the average commercial life of a rubber crop
is about 30 years (5 years immaturity, 20-25 years yielding;
Broughton, 1977), about one million hectares (i.e. one sixth of
the area sown to rubber) might contain a substantial amount of
centro. Where legume cover crops are sown, trees reach a tappable
size up to 12 months earlier (Watson, 1963), and the residual benefits may persist for the entire commercial lifetime of the rubber
crop (Broughton, 1977). These benefits are the result of reduced
soil erosion, better soil physical structure, fixation of atmospheric nitrogen, conservation and slow release of plant nutrients;

and improved rooting of the rubber trees, perhaps because of my-corrhizal root associations. Centro thus has an economically sig-nificant role in plantation agriculture.

Centro is cultivated as a pasture legume in most wet tropi-cal areas of the world (humid semi-hot equatorial and humid semi-hot trpical climates of Papadakis, 1966). In Australia there are about 20,000 hectares of centro-based pastures, mainly on the north Queensland coast, and this area is slowly expanding. Al-though about 5000 hectares have been sown during the last ten years, the net increase would be considerably less than this be-cause of losses due to overgrazing, inadequate fertilizer applica-tion, and ploughing of pastures for cropping. The area of centro-based pastures in other countries is not known. Extensive natural pastures containing appreciable amounts of centro occur in Venezu-ela and Central America, and large areas are being sown in Brazil.

Climate, Soils, and Fertilizers

The preferred annual rainfall for centro is 1500 mm or more, but centro persists in lower rainfall areas in Africa (800-1500 mm). It has some ability to tolerate waterlogging and will sur-vive a 3-4 month dry season, but is not adapted to prolonged drought. It is intolerant of low temperatures, growing very poor-ly when night temperatures fall below 15°C. Frosts of -3°C cause substantial leaf death, but plants may survive if growing points near the ground are protected. Thus, centro is a plant of the tropics, not well suited to the subtropics.

Soils of moderate fertility are preferred. Acceptable growth may be obtained on relatively acid, infertile soils with regular fertilizer application, and even on some very light sandy soils if the rainfall and fertilizer inputs are adequate. The acceptable pH range is from 4.5 (where extractable soil aluminum is less than 0.2 milliquivalents per 100 grams of soil) to 8.0, but nodulation is poor towards the extremes of the range and the optimum pH is about 5.5-6.0.

As responses to applied fertilizer depend on the nutrient status of the soil, there is little point in listing the require-ments for any given site. However, some generalizations can be made. Responses to phosphorus are most commonly obtained. Centro is usually considered to require higher levels of available soil phosphate than stylo (Stylosanthes guianensis) for maximum growth. Phosphorus uptake is greatly enhanced by mycorrhizal root develop-ment (Mosse et al., 1973, 1976). Centro also responds commonly to sulfur and calcium applications, and to the trace element molyb-denum. Less frequently, reponses to potassium and trace elements such as zinc, copper, and boron are obtained. There are frequent reports of the beneficial effects of liming, and these may be due to the correction of calcium deficiency, or more commonly to an increase in soil pH leading to increased molybdenum availability and a reduction in the toxic effects of aluminum and possibly man-ganese. Mineral nutrient deficiency symptoms have been described by Shorrocks (1964) and by Souto and Franco (1972).

In rubber plantations, typical fertilizer recommendations are 50 kg rock phosphate per hectare every 3-4 months for 12-18 months, followed by 100-200 kg per hectare each year while the cover crop persists (Anon., 1954, 1958). In Australian pastures, superphosphate is commonly recommended at rates equivalent to 100-200 kg per hectare per year, with additional fertilizers and trace elements to suit individual soil types (Teitzel et al., 1978). Elsewhere, experimental pastures have most often been fertilized with superphosphate at rates of 0-250 kg per hectare per year.

Establishment

Centro spreads from runners which root at the nodes, but new cover crops and pastures are usually established from seed. Seedling vigor is poor and establishment failures sometimes occur, particularly in pastures. Some seedbed preparation is usually necessary. Cover crops are drilled in rows, commonly using a 2:2:1 mixture of centro:calopo:puero. In pastures, centro seeding rates are 2-4 kg per hectare. Pastures may sometimes be established by broadcasting seeds in the cold ash after burning, at rates of 6-8 kg per hectare, or by oversowing into the existing poor grass sward. Centro is noted for having a high proportion of hard seeds, and hot water treatment or mechanical scarification of seed is usually necessary. Because of moderately specific Rhizobium requirement, inoculation of seed prior to sowing is recommended. Seed is relatively cheap. Recent commercial seed crops in Australia have yielded 70-360 kg per hectare with machine harvesting, and seed yields of up to 600 kg per hectare have been obtained elsewhere with hand picking.

Pests

Centro is susceptible to some serious pests and diseases (Sonoda and Lenne, 1979). In Malaysia, occasional outbreaks of the ladybird insect (Epilachna indica) may cause severe damage to the leaves, resulting in complete defoliation if insecticides are not applied. Flies (Melanagromyza centrosematis), caterpillars, grasshoppers, slugs, and snails are less important pests, and the fungal diseases Sporodesmium bakeri and Rhizoctonia solani (web blight) may cause minor damage. Web blight has also been reported from Florida (Sonoda et al., 1971). In Australia, Colombia, and Florida, leaf spot (Cercospora sp.) and red spider (Tetranychus sp.) attack the foliage, and in several countries Centrosema mosaic virus causes stunting and death of plants. Anthracnose (Colletotrichum sp.) has been observed in Florida (Lenne and Sonoda, 1978), and is a significant problem in Bolivia (Sonoda and Lenne, 1979) and Brazil. Other fungal pathogens are listed by Sonoda and Lenne (1979).

Grass-Legume Mixtures

Centro is compatible with a wide range of pasture grasses. In several countries, centro/guinea grass (Panicum maximum) mixtures have proved very successful. In Brazil, Malaysia, West Africa, and South-eastern U.S.A., pangola grass (Digitaria decumbens) has been sown with centro, but this mixture is not very successful in Australia. Other compatible grasses according to Bogdan (1977), Skerman (1977), and other authors include Cynodon plectostachyum (giant stargrass), Andropogon gayanus and Chloris gayana (Nigeria); Hyparrhenia rufa and Cenchrus ciliaris (East Africa); Heteropogon contortus and Pennisetum polystachyon (India); Melinis minutiflora, Panicum coloratum, and Pennisetum clandestinum (Australia); Pennisetum purpureum, Ischaemum indicum, Brachiaria humidicola, B. mutica, and B. decumbens (Pacific region); and Paspalum notatum and Setaria anceps (U.S.A.)

Established pastures often contain surprisingly little centro, but it is difficult to quantify this statement because there are remarkably few published records of botanical composiion of grazed centro-based pastures. Also the proportion of centro depends on seasonal effects (Bowen, 1959; Allen and Cowdry, 1961), stocking rate (Eng et al., 1978) and the adequacy of soil phosphorus and other nutrients, so variation from time to time may be expected. However, in grazing trials in Uganda (Horrell and Newhouse, 1966; Stobbs, 1969a), Australia (Harding and Cameron, 1972; Mellor, et al. 1973; Middleton et al., 1975) and Malaysia (Eng et al., 1978) centro has often contributed only 5-15 percent of the available herbage. It is not known to what extent these figures are biased by selective grazing, but deficiencies of phosphorus or molybdenum are probably major causes of low legume content. Where adequate phosphorus has been supplied, 30-50 percent legume has been obtained under grazing (Moore, 1962; McIlroy, 1962; Horrell and Newhouse, 1966; Grof and Harding, 1970b). Overgrazing in the dry season and trampling damage in the wet season are other likely causes of low legume yield. In cutting trials, the proportion of centro has varied even more widely.

Nitrogen Fixation

The amount of nitrogen fixed by centro depends on satisfactory nodulation, growth and persistence of the legume. Nodulation failures may be due to faulty inoculation techniques (e.g. use of ineffective strains of Rhizobium), or mineral nutrient deficiencies or toxicities, especially in very acid soils.

In pot trials and small plot experiments, estimates of nitrogen fixation have ranged from 90 to 400 kg N per hectare per year (Watson, 1957; Bowen, 1959; Whitney et al., 1967; Odu, et al. 1971). Fixation in grazed pastures is more difficult to measure and various techniques have been used. Under grazing, estimates of 80-280 kg N per hectare per year have been obtained (Moore, 1962; Bruce, 1965, 1967; Horrell and Newhouse, 1966; Stobbs, 1969b). In cutting trials, the proportion of centro has varied even more widely. Judging from these figures, in grazed pastures containing 5-15 percent legume, fixation would be unlikely to exceed 100 kg N per hectare per year. Where centro is grown alone as a cover crop in tree plantations, or where it contributes 30-50 percent of the herbage in grazed pastures, its fixation rate may approach the higher values recorded experimentally. However, the maximum fixation rate of cover crops containing centro, calopo, and puero is believed to be about 200 kg N per hectare per year (Broughton, 1977).

Herbage Quality

The chemical composition and herbage quality of centro is very well documented. Nitrogen content of leafy herbage is typically 3-4 percent (17-25 percent crude protein). Concentrations of several other elements have been tabulated by Wilson and Lansbury (1958), Moore (1962), Shorrocks (1964), and other authors. Selenium levels of 0.1 to 0.8 ppm were reported by Long and Marshall (1973). The crude fiber content is 25-40 percent (Oyenuga, 1957; Wilson and Lansbury, 1958; Miller and Rains, 1963), and the crude fiber digestibility is only 35-50 percent (Reyes, 1955; Wilson and Lansbury, 1958). Consequently, the digestibility of dry matter and organic matter is rather low, ranging from about 45 to 65 percent and commonly only 50-55 percent (Wilson and Lansbury, 1958; Miller and Rains, 1963; Johri et al., 1970; Reid et al., 1973). In one comparative experiment centro was slightly less digestible than siratro (Macroptilium atropurpureum) and stylo (Stylosanthes guianensis) but more digestible than Greenleaf desmodium (Desmodium intortum) at the same stage of growth (Reid et al., 1973). Some caution is necessary in interpreting these estimates, which have been obtained from cut herbage. In practice, the digestibility of the herbage actually eaten by grazing animals would be higher than these figures suggest because of preferential grazing of the more digestible leaf fraction (Stobbs, 1975).

Animal Production

Very high levels of animal production have been obtained from well-managed centro pastures. Some published data, chosen to represent a range of locations and pasture mixtures, are summarized in Table 2.4. The greatest animal production so far recorded (928 kg live weight gain per hectare per year) was obtained on a young centro/guinea grass pasture at Utchee Creek, Australia (Mellor et al., 1973). This is the highest recorded animal production from any tropical legume-based pasture in the absence of nitrogenous fertilizer. The levels more commonly obtained (Table 2.4) are of the order of 300-600 kg per hectare per year, and these are clearly well below the potential productivity of the system. The most common limitations to productivity are climatic (low rainfall, low temperatures) and nutritional (inadequate soil fertility, failure of the Rhizobium symbiosis). Reduced animal production may also result from undergrazing or overgrazing and, in some cases, from genetic limitations of the grazing animals, however, even though

TABLE 2.4
Animal Production from Pastures Containing <u>C. pubescens</u>

Location	Ibadan, NIGERIA	(a, b) Ngetta, (c, d) Serere, UGANDA	Utchee Creek, AUSTRALIA	Rio de Janeiro, BRAZIL
Mean annual rainfall (mm)	1200	1400	3250	1300
Associated grass	<u>Cynodon plectostachyum</u>	<u>Hyparrhenia rufa</u>	<u>Panicum maximum</u>	<u>Digitaria decumbens</u>
Superphosphate applied (kg ha^{-1} year^{-1})	(a) 190 (mean) (b) nil	(a) nil (c) 250 (b) 250 (d) nil	(a) nil? (b) 250 (1 yr only) (c) nil or very little	150 (mean)
Stocking rate (beasts ha^{-1})	(a) 4.2, 6.7 (b) 3.2-4.2	(a) 3.5 (c) 4.8 (b) 3.5 (d) 5.9	(a) 4.2 (b) 4.9 (wet season) 2.5 (dry season) (c) 2.7	2.4 (mean)
Liveweight gain (kg ha^{-1} year^{-1})	(a) 300-370 (b) 260	(a) 345 (c) 500 (b) 420 (d) 670	(a) 470 (b) 750 (c) 580	410
Liveweight gain (kg beast^{-1} day^{-1})	(a) 0.18-0.23 (b) 0.21	(a) 0.27 (c) 0.29 (b) 0.33 (d) 0.31	(a) 0.31 (b) 0.56 (c) 0.59	0.48
References	(a) Okorie et al., 1965 (b) Oyenuga & Olubajo, 1966	(a, b) Stobbs, 1969b (c) Stobbs, 1969c (d) Stobbs, 1969d	(a) Grof & Harding, 1970b (b) Mellor et al., 1973 (c) Mellor & Round, 1974	Aronovich et al., 1970

animal production may be well below the potential level, it is usually much greater than that achieved on unfertilized pure grass pastures.

OTHER CENTROSEMA SPECIES

C. sp. aff. pubescens cv. Belalto

Material similar to cv. Belalto has a restricted natural distribution, mainly in Central America (approximately 9°-21°N latitude), and especially in humid upland areas (600-1000 m) of Mexico and Costa Rica (humid tierra templada climates of Papadakis, 1966). Belalto (previously Q8333) is the only commercial cultivar and was collected on the San Jose Plateau (elevation 600 m, annual rainfall 2400 mm) in Costa Rica (Grof and Harding, 1970a). The soil at the collection site was derived from volcanic ash and was highly fertile. Belalto is similar to C. pubescens and experience with other material of this type is negligible.

Belalto is rather more prostrate than the commercial forms of C. pubescens, and differs in several obvious vegetative or floral characteristics (Harding and Cameron, 1972). Its flowers are more purple, its seeds are lighter in color, its leaflets more rounded, and growing points (including immature leaves) often have a reddish-purple pigmentation. It has greater stoloniferous root development than commercial C. pubescens, and will grow at lower temperatures (Grof and Harding, 1970a). It is more resistant than C. pubescens to Cercospora leaf spot and red spider (Tetranychus sp.).

In Australia, Belalto has proved to be a very successful legume on the wet tropical coast of Queensland near South Johnstone (lat. 17°S, annual rainfall 3400 mm; humid semi-hot tropical climate of Papadakis, 1966) on well fertilized basaltic soils. (It is less successful on nearby less fertile granitic soils where the rainfall is considerably lower). In this environment it associates well with guinea grass (Panicum maximum), and resists weed invasion better than common centro (Grof and Harding, 1970a; Harding and Cameron, 1972). In small plot trials at several locations in southern Queensland (lat. 25-29°S, rainfall 700-1000 mm; various subtropical climates according to Papadakis, 1966) on a range of soil types it has grown poorly or failed to persist during dry conditions. However at Beerwah (27°S) where the rainfall is higher (1600 mm; cool humid tropical climate of Papadakis, 1966) Belalto has grown and persisted well in both cutting and grazing trials.

A seeding rate of 5.5 kg/ha is recommended (Teitzel et al., 1974). Good seedbed preparation is advised, but successful establishment has been obtained by sod-seeding Belalto into a legume deficient guinea grass pasture (Harding, 1972). Inoculation with the commercial C. pubescens strain of Rhizobium results in effective nodulation. Nitrogen fixation capability has not been measured. Fertilizer recommendations in Queensland, Australia are the same as those for C. pubescens.

Once established, Belalto survives well under grazing. This has been atttributed partly to its strongly stoloniferous growth (Harding and Cameron, 1972). Animal production under grazing is currently being measured at Utchee Creek. In this trial, Belalto/ guinea grass pastures have persisted for 5 years at stocking rates of 3.7 beasts/ha during the wet season and 2.5 beasts/ha during the drier winter. In one period of 192 days during the 1975 wet season, these pastures produced 508 kg liveweight gain per hectare.

C. pascuorum

C. pascuorum is a self-regenerating annual species, native to semi-arid regions with a long dry season in the South and Central American tropics (hot tropical and semi-arid tropical climates of Papadakis, 1966). Most herbarium species and living accessions have been collected in northeastern Brazil, but the species also occurs in dry parts of Guyana, Venezuela, and Ecuador, and in isolated pockets in several Central American countries. There are several areas (e.g. the Guajira Peninsula in Colombia) which are climatically suitable but which have not yet been studied floristically, so the list of confirmed locations may eventually need some revision. There are reports in the literature of C. pascuorum occurring in southern Brazil, Argentina, Paraguay, and elsewhere, but in all of the cases we have investigated these reports have resulted from incorrect classification, and the extremes of the latitudinal range appear to be 16°N (Mexico) and 10-12°S (Brazil). Within these latitudes C. pascuorum has usually been collected from areas receiving less than 1000 mm annual rainfall. It has been found in higher rainfall areas in Venezuela (1000-1500 mm); however, where there is a regular, severe and prolonged dry season (4 to 6 months with less than 50 mm of rainfall).

Within these climatic boundaries the species is found on a diversity of soil types, with a pH range from 5 to 8.5 and a texture range from sand to heavy clay. The performance of individual accessions on a range of soil types in Australia, and in glasshouse experiments where the pH has been varied by applying lime, suggests that a good deal of edaphic adaptation probably exists within the species. It is not yet known whether this includes variation in phosphorus requirement, as the nutrient requirements and fertilizer responses of C. pascuorum have not been studied.

The first collections for agronomic evaluation of this species were made by R.J. Williams in 1965. Assessment in plant introduction nurseries commenced in 1966. Within Australia, these or subsequent accessions have been evaluated in small plots at many sites in Queensland, the Northern Territory, and the Kimberley region of Western Australia (Figure 2.2). C. pascuorum appears to be best suited to the monsoonal tropical climates of the 'Top End' of the Northern Territory (e.g. Katherine, lat. 14°S) and the Kimberleys (e.g. Kununurra, lat. 15°S), which correspond to the hot tropical and semi-arid tropical climates of Papadakis (1966). Here its annual habit allows it to escape drought stress and regenerate during successive wet seasons. Near Townsville (lat. 19°S; semi-arid monsoon subtropical and hot semi-tropical climate of Papadakis, 1966) it has persisted for several years but is clearly not so well adapted, while on the wet tropical coast (South Johnstone, lat. 17°S; humid semi-hot tropical climate of Papadakis, 1966) it does not persist. In the latitudinal range 23-28°S, most accessions flower too late to enable seed to be set before the winter. The earliest flowering accessions do sometimes set seed, however, and several accessions are able to perennate through a very mild winter and flower in the spring. At these subtropical latitudes, even if the soil is moist the species does not grow well during the spring because temperatures are too low, and competition from other species is too great. Thus, the Australian experience is consistent with its native distribution, and suggests that its area of potential use may be restricted to the seasonally arid tropics.

Differences in flowering date of more than 5 weeks between accessions have been observed (Clements and Williams, 1980). Further plant collection may reveal a much greater range of flowering dates and thus allow plant breeders and agronomists to select lines adapted to environments with shorter or longer growing seasons. Flowering time is strongly inherited and hybrids usually flower at dates intermediate between those of the parent accessions. At Katherine, N.T. Australia, accessions with near-optimum flowering dates have yielded more than 800 kg of seed per hectare in small plot trials conducted by W.H. Winter. All accessions so far examined have 100 percent hard seeds immediately after seed dispersal. A range of accessions is currently being screened to establish whether there is genetic variation in the rate of breakdown of hardseededness.

Although its annual habit is probably the main drought avoidance mechanism of C. pascuorum, the species can tolerate quite severe moisture stress during the vegetative phase. In both the Australian subtropics and tropics, if drought prevents flowering, some plants will perennate. It has been suggested that the characteristically narrow leaflets, and the manner in which the leaflets are pointed towards the sun (parahelionasty), enable the plants to keep their leaves cool while moisture is conserved (Clements and Williams, 1980). During drought, newly-produced leaflets become steadily smaller, narrower and more hairy, and parahelionasty becomes more pronounced. On the other hand, where moisture is plentiful many ecotypes may show striking stoloniferous root development, particularly on sandy soils. This is encouraged by the prostrate growth habit of the species when competition is absent. However, C. pascuorum can adopt a twining, scrambling habit in pastures, and this enables it to compete with associated grasses.

C. pascuorum is susceptible to a root knot nematode (Meloidogyne sp.), but this appears to be a problem only in experiments involving widely spaced plants, where deaths of individual plants are obvious. The symptoms and severity of infection vary among accessions, but it appears that none of the accessions in the existing collection is completely resistant. Individual plants

Figure 2.2. C. pascuorum growing at Katherine, Northern Territory, Australia.

are attacked by the legume little-leaf virus but again this is
only a problem in experiments with spaced plants. Thus there are
no serious disease problems at this stage.

The Rhizobium specificity of C. pascuorum is not known. In
Australia, all seeds or seedlings have been inoculated routinely
using the commercial Centrosema inoculum, which has resulted in
excellent nodulation with all accessions. The amount of nitrogen
fixed by the species has not been studied and its fertilizer requirements are unknown. In all experiments so far, the known soil
nutrient deficiencies have been corrected.

C. pascuorum has not yet been adequately tested under grazing
but some information is available from experimental sowings at
Katherine N.T. Australia. These early observations by W.H. Winter
suggest that the species should persist well once it is established. Because it is an annual, seed production in situ is essential.

Figure 2.3. C. virginianum CPI51038, collected from northern
Argentina.

Information on the performance of C. pascuorum in other countries is meager. Preliminary results from Khon Kaen in Thailand
are favorable, but only moderate success has been obtained in
Florida. At Serdang, Malaysia, results have been very disappoint-
ing.

C. virginianum

C. virginianum (Figure 2.3) has the most extensive natural
distribution of all the Centrosema species, occurring more or less
continuously from latitude 35°S (Uruguay and Argentina) to 40°N
(eastern U.S.A.). Within this extensive area it has been most
commonly collected from climates which are subhumid (500-1000 mm
annual rainfall) and subtropical by virtue of latitude or altitude. It has been occasionally collected from more tropical climates, including the semi-arid regions of Brazil, but is rarely
found in the wet tropics. Its distribution thus spans a great
range of climates according to Papadakis (1966). Because of its
wide distribution it possesses a number of forms or varieties,

some of which have been erroneously given separate species names
(Clements and Williams, 1980). The forms studied so far have
proved to be quite cross-compatible and should not be recognized
at other than the subspecific level.

C. virginianum is a perennial, twining plant. It is rather
like C. pubescens in its gross morphology, but generally less ro-
bust, with narrower leaflets, more continuous flowering and seeding, and little or no stoloniferous root development. It can be
usually distinguished from C. pubescens by its calyx characteristics (all 5 calyx lobes are usually much longer than the tube, and
more nearly of equal length than in C. pubescens) and its seeds,
which are usually cylindrical in shape. With experience, the two
species can be separated on many other vegetative and floral characteristics. Despite this they have often been confused, partly
because the calyx characteristics are not always distinctively
different.

C. virginianum is genetically the most variable of all the
Centrosema species (Figure 2.4), particularly for agronomic characters such as vigor, flowering behavior, frost resistance (winter
survival), and drought resistance (Clements, 1977a; Clements and
Ludlow, 1977). This variation, coupled with the wide distribution
of the species, suggests that it may be a useful pasture legume in
the subtropics. Although a representative collection of introduced lines has been evaluated in Australia, none has proved suitable
for commercial use (Clements, 1977a). By plant breeding, however,
it may be possible to recombine the vigor of the tropical forms
with the prolific seeding and frost survival of the subtropical
ecotypes, and thus produce commercially acceptable varieties
(Clements, 1977b).

C. virginianum has been collected from a wide range of soils,
but its nutrient requirements and tolerances have not been studied. It is more drought resistant than C. pubescens but less tolerant of waterlogging. In Australia it has been grown successfully both on moderately acid, sandy soils (pH 5.3-5.5), and alkaline
clay soils (surface pH 6.5, increasing to pH 9 with depth). Superphosphate has been applied at low levels (0-100 kg per hectare
per year) after correcting any initial gross deficiencies of phosphorus and other nutrients. As a routine procedure, seeds or
seedlings have been inoculated with the commercial centro strain
of Rhizobium, and effective nodulation is always obtained. Seeding rates of 2-3 kg per hectare have given satisfactory establishment. Associated grasses have included Cenchrus ciliaris at Narayen Research Station, Queensland, Australia (lat. 26°S, annual
rainfall 720 mm) and Setaria anceps at Beerwah Research Station,
Queensland, Australia (lat. 27°S, annual rainfall 1600 mm). C.
virginianum appears to have no serious pest or disease problems.

Laboratory tests have shown that in vitro digestibility of
leaf material (54-59 percent) is similar to siratro (Macroptilium
atropurpureum; 56 percent), but stem digestibility is somewhat
lower (Clements, 1977a). The herbage is accepted readily by cattle. Persistence and productivity of bred lines under grazing is
being assessed at Beerwah and Narayen, but these trials are at a
very early stage.

Figure 2.4. Pods and seeds of C. virginianum (a, b, c) and C.
pascuorum (d, e, f) showing differences in size and
shape.

C. schottii

C. schottii is a rather coarse, scrambling, broad-leafed species, similar in many respects to C. plumieri, with which it is
often confused. Figure 37 of Skerman (1977) is in fact C. schottii and not C. plumieri as indicated in the caption. The two
species may be separated on many counts, particularly by the shape
of the terminal leaflet (laterally lobed more strongly in the case
of C. schottii), flower color (mauve for C. schottii; white with
red markings, or more rarely red for C. plumieri) and pod characteristics (C. schottii has much narrower pods, typically pale
brown to yellow when ripe). The seeds of C. schottii are smaller
and often have a dull, fuzzy appearance when first harvested,
quite unlike the hard, shiny seeds of C. plumieri.
In nature, C. schottii has a restricted and disjunct distribution, and geographically isolated collections have incorrectly
been given several different names. Thus in Brazil it has been

called C. macranthum; in Haiti, C. haitiense, in Cuba, C. lobatum; and in Argentina, C. kermesi. It appears to be most common on the Yucatan Peninsula of Mexico (550-1500 mm annual rainfall) and the Pacific coast of Ecuador. In these locations the soils tend to be strongly alkaline, and perhaps it may be specifically adapted to neutral or alkaline soils (Clements and Williams, 1980) rather than to any particular climatic regime.

C. schottii has been grown in small plot trials at several locations in Australia. Its best performance so far has been at Katherine, N.T. A thorough evaluation of the full range of variation within the species is needed.

C. brasilianum

C. brasilianum has a wide natural distribution in the South American tropics, but it is particularly common in semi-arid northeastern Brazil (hot tropical and semi-arid climates of Papadakis, 1966). Although this species has also been found in the humid Amazonian region, its greatest potential value lies in the drought resistance of the dryland ecotypes. The present living collections contain only the most common form of C. brasilianum, which has rather narrow leaflets and slender stems. Broad-leafed forms are known to exist, and there are several closely related species which have not been collected for evaluation.

The accessions in the Australian collection do not nodulate effectively with the commercial centro strain of Rhizobium, which has limited the assessment of their agricultural value. Also there is some evidence that the species is not well adapted to the subtropics, but it will be necessary to collect and evaluate material known to exist near the Tropic of Capricorn (Chaco region) before the area of adaptation can be defined.

The first real indication of the productivity and drought resistance of C. brasilianum was obtained by W.H. Winter in trials at Katherine, N.T. Australia and by R.L. Burt in northern Queensland, where nodulation with native rhizobia occurred. This experience led to a more systematic evaluation of the collection, and to the preliminary assessment of C. brasilianum under grazing. An effective Rhizobium strain has now been isolated from Katherine N.T., Australia.

C. plumieri

This well-known species is a scrambling perennial with large leaves, rather thick stems, and large, attractive flowers. Because of its vigor, it was at one time grown as a cover crop on plantations in Java and the Malay Peninsula (Burkhill, 1935). Although it grows well in the wet tropics (e.g. humid, semi-hot, tropical climates of Papadakis, 1966), its wide natural distribution extends into sub-humid climates. It is particularly common in the Caribbean region. As with C. pubescens, the cultivated forms are relatively homogenous, but there is considerable diversity among the small number of accessions that have been obtained by field collection in South and Central America (e.g. characteristics such as flower color, seed size and color, flowering date, and plant vigor). Until a representative collection is obtained and evaluated, our understanding of the adaptive range and agricultural potential of this species will remain incomplete.

C. plumieri has shown some promise as a pasture plant. When it is allowed to grow without defoliation it gives high herbage yields (e.g. Grof and Harding, 1970a), and its digestibility is superior to C. pubescens (Reid et al., 1973). However, the genetically restricted range of material that has been tested does not thrive under close cutting (Grof and Harding, 1970a) or grazing.

GENETIC IMPROVEMENT

Each of the Centrosema species we have discussed is genetically variable. As a starting point, improvement programs should aim to exploit this naturally occurring variability. Only if the desired combination of characters does not exist within the available genetic resource should plant breeding be considered. Plant breeding is long-term in nature, and requires a high level of commitment by breeders and the institutions that employ them. Without a substantial input in terms of time, money, and facilities, plant breeding is unlikely to succeed.

Screening of natural ecotypes of Centrosema species has been conducted or is in progress in several countries (Australia, Colombia, subtropical U.S.A., Brazil, Thailand, and Malaysia). The information obtained from this regional evaluation is of fundamental importance to plant breeders, agronomists, and taxonomists. A small proportion of the data has been published (e.g. Grof and Harding, 1970a; Hymowitz, 1971; Harding, 1972; Clements, 1977a; Kretschmer, 1977); however, most of the information remains unpublished and is probably unpublishable in scientific journals in its present form. This is understandable, because with a continuing inflow of large numbers of accessions a contemporary evaluation of the whole collection becomes impossible, and evaluation proceeds in a piecemeal fashion. Concurrent with the assembly of large collections in a few major centers, we believe an attempt should be made to collate the regional data in a form which can be made available to collaborating institutions. The current trend towards computerized data storage and retrieval systems will be of little value unless these systems are used to produce catalogues of origin and performance information.

All Centrosema cultivars currently in commercial use have arisen from naturally occurring material. They include the common strains of C. pubescens and its Brazilian cultivar Deodoro, which was collected near Rio de Janeiro according to Souto and de Lucas (1972), and the C. sp. aff. pubescens cultivar Belalto (Grof and Harding, 1970a). There is little doubt that more varieties will be released as more material is collected and evaluated.

Most plant breeding work has been done with C. pubescens. In Australia, and later in Colombia, Grof (1970) has bred high yielding, stoloniferous lines which are resistant to Cercospora leaf spot and Centrosema mosaic virus. A new variety derived from this bred material is presently being tested under grazing in Columbia, in association with Andropogon gayanus. In Brazil, Serpa (1972, 1977) has bred lines with more rapid seedling growth and early nitrogen fixation, and is now breeding for resistance to anthracnose. Serpa's material appears to be derived from crosses among C. pubescens ecotypes, rather than interspecific hybridization as reported. In Colombia, Hutton is attempting to breed for tolerance to high levels of soil aluminum, a significant objective but one which may be very difficult to achieve. According to Schultze-Kraft and Giacometti (1979), Centrosema is not a genus of indicator plants for infertile acid soils. Although some species do occur on Oxisols in the Colombian Llanos (notably C. angustifolium, C. macrocarpum, and C. venosum) the occurrence of C. pubescens on such soils is rare, though it is found on adjacent gallery forest soils which may be more fertile.

There is ample scope for further breeding work with C. pubescens. Breeders so far have concentrated on some of the important deficiencies of the species, but many other objectives could be justified. These include drought resistance, cold resistance, better herbage quality, and more efficient phosphorus utilization. Improvement in these characters would have a major effect on the adaptability and economic value of C. pubescens.

In Australia, Clements (1977b) is breeding C. virginianum for subhumid subtropical conditions and C. pascuorum (Clements, 1979) for the seasonally arid tropics. These programs are at an intermediate stage. Some crosses have also been made within C. schottii, C. brasilianum, and C. plumieri but there has been no attempt to select superior plants from the progeny.

Much more work is needed to clarify evolutionary and genetic relationships between species. This is particularly true for the group of species closely related to C. pubescens, i.e. C. grandiflorum, C. macrocarpum, C. acutifolium, C. capitatum, and C. magnificum. If these species could be successfully hybridized with C. pubescens, the range of adaptation for C. pubescens could be greatly extended. Although these particular relationships are the most agriculturally significant, several other taxonomic problems will only be solved by cytogenetic and biochemical studies when living material of rare species is obtained.

CONCLUSIONS

Two general conclusions emerge from this review:

(1) The economic potential of the genus as a whole is poorly understood. The cultivated forms of C. pubescens, C. sp. aff. pubescens cv. Belalto and C. plumieri only represent a fraction of the diversity within Centrosema. The remaining diversity is almost untapped and largely unexplored, but there are already indications that several other species may have agricultural potential. Further plant collection and evaluation should receive high priority in future research. Only when representative collections are available will it be possible to carry out the agricultural,

genetic, and taxonomic research essential both to an understanding of the pattern of adaptation of the genus as a whole and to the recognition and development of economically important species.

(2) The performance of C. pubescens in pastures has been shown experimentally to be potentially very high, but the level of performance actually achieved is often much lower than it should be. In seeking to understand and correct the limitations to production, information is scarcely past the pioneering stage. Whether it is used as a pasture plant or a cover crop, a major requirement of C. pubescens is that it must fix large quantities of atmospheric nitrogen, and it will only do this if it produces large quantities of herbage. Plant growth and nitrogen fixation are inextricably linked. There is need to know much more about the legume/Rhizobium symbiosis in centro, but studies are also required on the particular eccentricities of the plant including its reaction to pests and diseases and to grazing pressure. In particular there appear to be problems of mineral nutrition in pastures which are poorly understood. Perhaps studies on the comparative nutrient requirements of centro and other successful pasture legumes will lead to progress in this area.

REFERENCES

Ahlgren, G.H. 1959. Development of grasslands in the western region of Nigeria. Ministry of Agriculture and Natural Resources, Ibadan, Nigeria.

Allen, G.H. and Cowdry, W.A.R. 1961. Yields from irrigated pastures in the Burdekin. Queensland Agricultural Journal 87: 207-213.

Anonymous 1954. Establishing a legume cover. Planters' Bulletin No. 14: 86-94.

Anonymous 1958. Establishment and maintenance of legume covers. Planters' Bulletin No. 39: 129-133.

Aronovich, S., Serpa, A., and Ribeiro, H. 1970. Effect of nitrogen fertilizer and legume upon beef production of pangolagrass pasture. Proceedings of the XI International Grassland Congress, Surfers Paradise, 1970: 796-800.

Barbosa-Fevereiro, V.P. 1977. Centrosema (A.P. de Candolle) Bentham do Brasil - Leguminosae-Faboideae. Rodriguesia XXIX (42): 159-219.

Bentham, G. 1837. Commentationes de leguminosarum generibus. Wien.

Bogdan, A.V. 1977. Tropical Pasture and Fodder Plants. Longman, London: 330-335.

Bowen, G.D. 1959. Field studies on nodulation and growth of Centrosema pubescens Benth. Queensland Journal of Agricultural Science 16: 253-265.

Broughton, W.J. 1977. Effect of various covers on soil fertility under Hevea brasiliensis Muell. Arg. and on growth of the tree. Agro-Ecosystems 3: 147-170.

Bruce, R.C. 1965. Effect of Centrosema pubescens Benth. on soil fertility in the humid tropics. Queensland Journal of Agricultural and Animal Sciences 22: 211-226.

Bruce, R.C. 1967. Tropical legumes lift soil nitrogen. Queensland Agricultural Journal 93: 221-226.

Burkill, I.H. 1935. A Dictionary of the Economic Products of the Malay Peninsula. Crown Agents for the Colonies, London; Vol. 1: 509.

Clements, R.J. 1977a. Agronomic variation in Centrosema virginianum in relation to its use as a sub-tropical pasture plant. Australian Journal of Experimental Agriculture and Animal Husbandry 17: 435-444.

Clements, R.J. 1977b. Plant breeding strategies for Centrosema virginianum. CSIRO Division of Tropical Crops and Pastures Annual Report 1975-76: 95-99.

Clements, R.J. 1979. Centrosema breeding. CSIRO, Division of Tropical Crops and Pastures Annual Report 1977-78: 51.

Clements, R.J. and Ludlow, M.M. 1977. Frost avoidance and frost resistance in Centrosema virginianum. Journal of Applied Ecology 14: 551-566.

Clements, R.J. and Williams, R.J. 1980. Genetic diversity in Centrosema. In "Advances in Legume Science". (eds.) R.J. Summerfield and A.H. Bunting. Royal Botanic Gardens, Kew. pp. 559-567.

Eng, P.K., Kerridge, P.C. and 't Mannetje, L. 1978. Effects of phosphorus and stocking rate on pasture and animal production from a guinea grass-legume pasture in Johore, Malaysia. I. Dry matter yields, botanical and chemical composition. Tropical Grasslands 12: 188-197.

Frahm-Leliveld, J.A. 1953. Some chromosome numbers in tropical leguminous plants. Euphytica 2: 46-48.

Grof, B. 1970. Interspecific hybridization in Centrosema: hybrids between C. brasilianum, C. virginianum, and C. pubescens. Queensland Journal of Agricultural and Animal Sciences 27: 385-390.

Grof, B. and Harding, W.A.T. 1970a. Yield attributes of some species and ecotypes of Centrosema in north Queensland. Queensland Journal of Agricultural and Animal Sciences 27: 237-243.

Grof, B. and Harding, W.A.T. 1970b. Dry matter yields and animal production of guinea grass (Panicum maximum) on the humid tropical coast of north Queensland. Tropical Grasslands 4: 85-95.

Harding, W.A.T. 1972. The contribution of plant introduction to pasture development in the wet tropics of Queensland. Tropical Grasslands 6: 191-199.

Harding, W.A.T. and Cameron, D.G. 1972. New pasture legumes for the wet tropics. Queensland Agricultural Journal 98: 394-406.

Horrell, C.R. and Newhouse, P.W. 1966. Yields of sown pastures in Uganda, as influenced by legumes and by fertilizers. Proceedings of the IX International Grassland Congress, Sao Paulo, 1965: 1133-1136.

Humphreys, L.R. 1974. A guide to better pastures for the tropics and sub-tropics. Third edition. Wright, Stephenson & Co. (Australia) Pty, Ltd., Melbourne: 56-57.

94

Hymowitz, T. 1971. Collection and evaluation of tropical and sub-tropical Brazilian forage legumes. _Tropical Agriculture (Trinidad)_ 48: 309-315.

Johri, P.N., Ahmad, N., and Jha, G.D. 1970. Chemical composition, digestibility and nutritive value of butterfly-pea (_Centrosema pubescens_ Benth.) at the pod stage. _Indian Journal of Agricultural Science_ 40: 33-35.

Kretschmer, A.E. 1977. Growth and adaptability of _Centrosema_ species in South Florida. _Soil and Crop Science Society of Florida Proceedings_ 36: 164-168.

Lenne, J.M. and Sonoda, R.M. 1978. _Colletotrichum_ spp. on tropical forage legumes. _Plant Disease Reporter_ 62: 813-817.

Long, M.I.E. and Marshall, B. 1973. The selenium status of pastures in Uganda. _Tropical Agriculture (Trinidad)_ 50: 121-128.

McIlroy, R.J. 1962. Grassland improvement and utilization in Nigeria. _Outlook on Agriculture_ 4: 174-179.

Mellor, W., Hibberd, M.J., and Grof, B. 1973. Beef cattle live-weight gains from mixed pastures of some guinea grasses and legumes on the wet tropical coast of Queensland. _Queensland Journal of Agricultural and Animal Sciences_ 30: 259-266.

Mellor, W. and Round, P.J. 1974. Performance of beef steers in the wet tropics of Queensland. _Queensland Journal of Agricultural and Animal Sciences_ 31: 213-220.

Middleton, C.H., Mellor, W., and McCosker, T.H. 1975. Agronomic limitations to pasture and animal performance in the wet tropics. Australian Conference on Tropical Pastures, Townsville, Queensland, May 1975, Vol. 1(c)25-1(c)29.

Miller, T.B. and Rains, A.B. 1963. The nutritive value and agronomic aspects of some fodders in Northern Nigeria. _Journal of the British Grassland Society_ 18: 158-167.

Moore, A.W. 1962. The influence of a legume on soil fertility under a grazed tropical pasture. _Empire Journal of Experimental Agriculture_ 30: 239-249.

Mosse, B., Hayman, D.S., and Arnold, D.J. 1973. Plant growth responses to vesicular-arbuscular mycorrhiza. V. Phosphate uptake by three plant species from P-deficient soils labelled with ^{32}P. _New Phytologist_ 72: 809-815.

Mosse, B., Powell, C.L., and Hayman, D.S. 1976. Plant growth responses to vesicular-arbuscular mycorrhiza. IX. Interactions between VA mycorrhiza, rock phosphate and symbiotic nitrogen fixation. _New Phytologist_ 76: 331-342.

Odu, C.T.I., Fayemi, A.A., and Ogunwale, J.A. 1971. Effect of pH on the growth, nodulation and nitrogen fixation of _Centrosema pubescens_ and _Stylosanthes gracilis_. _Journal of the Science of Food and Agriculture_ 22: 57-59.

Okorie, I.I., Hill, D.H., and McIlroy, R.J. 1965. The productivity and nutritive value of tropical grass/legume pastures rotationally grazed by N'Dama cattle at Ibadan, Nigeria. _Journal of Agricultural Science_ 64: 235-245.

Oyenuga, V.A. 1957. The composition and agricultural value of some grass species in Nigeria. _Empire Journal of Experimental Agriculture_ 25: 237-255.

Oyenuga, V.A. and Olubajo, F.O. 1966. Productivity and nutritive value of tropical pastures at Ibadan. Proceedings of the IX International Grassland Congress, Sao Paulo, 1965: 962-969.

Papadakis, J. 1966. Climates of the World and their Agricultural Potentialities. Author, Av. Cordoba 4564, Buenos Aires: 36: 99.

Reid, R.L., Post, A.J., Olsen, F.J., and Mugerwa, J.S. 1973. Studies on the nutritional quality of grasses and legumes in Uganda. I. Application of in vitro digestibility techniques to species and stage of growth effects. Tropical Agriculture (Trinidad) 50: 1-15.

Reyes, B. 1955. The digestibility of Centrosema pubescens and Pueraria javanica. The Philippine Agriculturist 39: 27-29.

Schofield, J.L. 1941. Introduced legumes in North Queensland. Queensland Agricultural Journal 56: 378-388.

Schultze-Kraft, R. and Giacometti, D.C. 1979. Genetic resources for forage legumes for the acid, infertile savannas of tropical America. In "Pasture Production in Acid Soils of the Tropics" (eds.) P.A. Sanchez and L.E. Tergas, CIAT, Cali, Colombia; pp. 55-64.

Serpa, A. 1972. Selecao precoce para nitrogenio total em Centrosema pubescens. Pesquisa Agropecuaria Brasileira, Serie Zootecnia 7: 29-31.

Serpa, A. 1977. Hibridacao interspecifica entre Centrosema pubescens e Centrosema virginianum. Pesquisa Agropecuaria Brasileira, Unico 12: 35-40.

Shorrocks, V.M. 1964. Mineral Deficiencies in Hevea and Associated Cover Plants. Rubber Research Institute of Malaysia, Kuala Lumpur.

Skerman, P.J. 1977. Tropical Forage Legumes. F.A.O. Plant Production and Protection Series No. 2. FAO Rome, pp. 244-258.

Sonoda, R.M., Kretschmer, A.E., and Brolmann, J.B. 1971. Web-blight of introduced forage legumes in Florida. Tropical Grasslands 5: 105-107.

Sonoda, R.M. and Lenne, J.M. 1979. Diseases of Centrosema spp. Fort Pierce (University of Florida) Agricultural Research Center Research Report RL-1979-3.

Souto, S.M. and Franco, A.A. 1972. Sintomalogia de deficiencia de macronutrientes em Centrosema pubescens e Phaseolus atropurpureus. Pesquisa Agropecuaria Brasileira, Serie Zootecnia 7: 23-27.

Souto, S.M. and de Lucas, E.D. 1972. Estabelecimento de leguminosas forrageiras tropicais. Pesquisa Agropecuaria Brasiliera, Serie Zootecnia 7: 33-38.

Stobbs, T.H. 1969a. The influence of inorganic fertilizers upon the adaptation, persistency and production of grass and grass/legume swards in Eastern Uganda. East African Agriculture and Forestry Journal 35: 112-117.

Stobbs, T.H. 1969b. The value of Centrosema pubescens (Benth.) for increasing animal production and improving soil fertility in Northern Uganda. East African Agriculture and Forestry Journal 35: 197-202.

Stobbs, T.H. 1969c. The use of liveweight-gain trials for pasture evaluation in the tropics. 3. The measurement of large pasture differences. Journal of the British Grassland Society 24: 177-183.

Stobbs, T.H. 1969d. The effect of grazing management upon pasture productivity in Uganda. 2. Grazing frequency. Tropical Agriculture (Trinidad) 46: 195-200.

Stobbs, T.H. 1975. Factors limiting the nutritional value of grazed tropical pastures for beef and milk production. Tropical Grasslands 9: 141-150.

Teitzel, J.K., Abbott, R.A., and Mellor, W. 1974. Beef cattle pastures in the wet tropics. 3. Pasture species. Queensland Agricultural Journal 100: 185-189.

Teitzel, J.K. and Burt, R.L. 1976. Centrosema pubescens in Australia. Tropical Grasslands 10: 5-14.

Teitzel, J.K. Standley, J., and Wilson, R.J. 1978. Maintenance fertilizer strategies for wet tropics pastures. Queensland Agricultural Journal 104: 126-130.

Toutain, B. 1973. Principales plantes fourrageres tropicales cultivees. Institut d'Elevage et de Medecine Veterinaire des Pays Tropicaux, Note de Synthese, No. 3.

Watson, G.A. 1957. Nitrogen fixation by Centrosema pubescens. Journal of the Rubber Research Institute of Malaya 15: 168-174.

Watson, G.A. 1963. Cover plants and tree growth, Part 1. The effect of leguminous and non-leguminous cover plants on the period of immaturity. Planters' Bulletin of the Rubber Research Institute of Malaya 68: 123-129.

Whitney, A.S., Kanehiro, Y., and Sherman, G.D. 1967. Nitrogen relationships of three tropical forage legumes in pure stands and in grass mixtures. Agronomy Journal 59: 47-50.

Wilson, A.S.B. and Lansbury, T.J. 1958. Centrosema pubescens: ground cover and forage crop in cleared rain forest in Ghana. Empire Journal of Experimental Agriculture 26: 351-364.

3
Desmodium

B. C. Imrie, R. M. Jones,
and P. C. Kerridge

THE POTENTIAL RESOURCE

The genus Desmodium contains mostly perennial herbs or sub-shrubs, less commonly shrubs, and rarely trees. Desmodium is distinguished by its jointed pods which are usually indehiscent (do not split open at maturity); a few species do have dehiscent pods. The genus has provided a number of species used as pasture and fodder crops, as ground cover and green manure, and in primitive medicinal preparations. Desmodiums are commonly known as tick clovers.

TAXONOMY

The following brief summary of the taxonomy of Desmodium follows the classification by Ohashi (1973). Ohashi placed Desmodium in the tribe Coronilleae, subtribe Desmodiinae, together with 24 other genera including the closely related Codariocalyx, Dicerma, Dendrolobium, Hegnera, Phyllodium, and Tadehagi. Other authors (e.g. Hutchinson, 1964) have elevated the subtribe to tribe rank, the Desmodieae.
Ohashi's classification based on Asiatic species, uses the broad genus concept of Desmodium to be "an assemblage of considerably heterogeneous species groups [that] cannot be divided into smaller genera due to the continuity in main characters." This concept appears to be shared by Schubert who has published widely on the taxonomy of American and African desmodiums but has not produced a comprehensive monograph. Schubert (1963) recognized three subgenera in the Mexican desmodiums and considered inflorescence and loment (segmented seed pod) characters the most useful for classifying species.
Because of the wide morphological variation in Desmodium and its close relatives, the continuity of taxonomic characters across species borders, and the opportunity to base classifications on different combinations of characters, the classification of the genus has received different approaches by taxonomists and appears somewhat confused. Estimates of the number of species have

varied, but it is most likely that there are between 350 and 450 species.

ORIGIN AND DISTRIBUTION

Of the seven genera recognized by Ohashi in subtribe Desmodiinae, five are restricted in their distribution to Southeast Asia, whereas Desmodium and Dendrolobium are more widely distributed. Because of the presence of related genera, and the degree of differentiation at the subgeneric and species levels, Southeast Asia is considered to be the center of origin and primary center of differentiation of Desmodium. Most species occur in an area incorporating India, Burma, Thailand, Indo-China, western and south-western China, and Malaysia. The second major center of diversity of Desmodium is in Mexico, with the Mexican species being distinct from those originating in Asia. In addition to these major centers, desmodiums are found in most tropical and subtropical parts of the world. They have been reported in the Floras of all continents except Europe. Distribution of species according to the Index Kewensis is approximately 160 in equatorial regions (0-10° lat.), 180 in tropical and subtropical regions (10-25° lat.), and 110 species in temperate regions (above 25° lat.).

Distribution of species with considerable agricultural potential is as follows:

D. adscendens - Asia (India, Sri Lanka, Thailand, Malaysia), Melanesia, Africa (West Africa, Zaire, Angola) Central America and the West Indies, and northern South America.

D. barbatum - Endemic to southern India but also found widely growing in Central and South America, and Africa (Rhodesia, Zambia, Malawi).

D. canum - Central America and South America as far south as Paraguay and Argentina.

D. discolor - Central America and northern South America.

D. distortum - Central America.

D. heterocarpon - Africa, Australia, Pacific Islands, and widespread throughout Asia.

D. heterophyllum - New Guinea, the Pacific Islands, and S.E. Asia.

D. intortum - Central America, West Indies, and northern South America.

D. sandwicense - Central America and northern South America.

D. uncinatum - Brazil and Bolivia.

ADAPTATION TO SOILS AND CLIMATE

The natural distribution of Desmodium, being predominantly in tropical and subtropical zones, provides the best indication of its climatic adaptation. Within equatorial regions, species of

Desmodium are found from sea level to 3000m whereas at higher latitudes they occur at lower altitudes. Desmodiums occur mostly in humid to sub-humid regions where the precipitation/evaporation ratio is high in most months of the year. They are therefore basically plants of warm, moist areas. Given these climatic restrictions, they are invariably found on acidic (pH less than 6.5) soils. Their habitat is usually open woodland or forest clearings or margins. They are found less commonly in savannahs. However, as may be expected in such a diverse and variable genus, there are species which are adapted to more extreme environments. For example, D. salicifolium grows well in swampy areas, D. uncinatum prefers moist lowlands, and D. hirtum is indigenous on saline soils in Senegal.

MORPHOLOGY

Habit

Of the subgenera of Desmodium; Ougeinia is a tree, Catenaria and Dollinera are shrubs, Hanslia is a sub-shrubby climber, and Podocarpium contains mostly herbs. Desmodium and Sagotia include both herbs and shrubs. However, even within a species there may be variation such as the range from prostrate herb to erect shrub encountered in D. barbatum.

Plant Parts

Leaflet number may vary from one to seven, with 1- and 3-foliate leaves being most common. Leaflet number can vary on a single plant as well as among species. Inflorescences are basically racemes with variation reflecting phylogenetic (evolutionary) relationships; more advanced species have simpler inflorescences than less advanced species. Flower color varies from white through pink and mauve to purple. Pods have from two to twelve segments, are mostly indehiscent, and covered with hairs. Seeds are generally smooth, slightly swollen, reniform (kidney shaped), transversely elliptic or depressed ovate. The hilum (point of attachment to the pod) is small.

Seedlings

The first two leaves above the cotyledon are borne opposed at the first node. They have simple laminae, no stipels, and are almost universally broadly ovate. The third and subsequent leaves are borne alternately and assume the characteristic form of adult leaves.

GENETICS

Chromosome numbers of approximately 10 percent of Desmodium species have been determined. In most species there are 2n = 22 chromosomes. Confirmed exceptions are 2n = 20 for the African species D. salicifolium and the American species D. painteri (Pritchard and Gould, 1964; Rotar and Urata, 1967). Chromosome numbers of related genera are mostly 2n = 22 but 2n = 20 has been reported in Codariocalyx (Pritchard and Gould, 1964). Flowering behavior and hybridization have been studied in a limited number of species. Flowering response to daylength varies with species and may be insensitive (D. sandwicense), long day (some strains of D. canum), or short day (D. intortum, D. uncinatum) (Rotar et al., 1967, Chow and Crowder, 1974). However, flowering of both short and long day plants will usually only occur at the end of the growing season under field conditions. Self-incompatability has not been reported in Desmodium and species appear to be predominantly self-pollinated. However, since fertilization and seed set may be improved by tripping (mechanical interference with a flower causing it to spring open and release a shower of pollen) of flowers there is ample opportunity for substantial out-crossing where bees are largely responsible for flower tripping. Outcrossing rates up to 30 percent in D. sandwicense and 8 percent in D. intortum have been measured.

The success of interspecific hybridization is dependent on how closely related the parent species are. D. intortum and D. sandwicense hybridize readily though failure of chromosome pairing has been noted (Hacker, 1968). Successful crosses have also been reported for all combinations between D. uncinatum, D. sandwicense, D. intortum. D. canum, D. aparines, D. sericophyllum, and D. pringlei (Chow and Crowder, 1972; Hutton and Gray, 1967). Validity of the above list may include either or both species. Viability and fertility of the hybrids derived from crosses listed above varied considerably.

The number of genetic studies published is not very large. Seed size (in D. canum, D. intortum, D. sandwicense), internode and raceme length (in D. intortum, D. sandwicense, D. uncinatum) and nodulating ability (D. intortum) have been reported to be polygenically controlled (Imrie, 1972a, 1975; Rotar and Chow, 1971; Hutton and Coote, 1972). Several morphological characters have been determined to be controlled by a single gene pair. These include flower color (colored dominant to near white in D. sandwicense and D. intortum), stem color (red or brown dominant to green in D. canum, D. sandwicense, D. intortum, D. uncinatum), leaf markings (silver midrib of D. canum, D. sandwicense, and D. uncinatum dominant to unmarked, and brown flecks in D. intortum dominant to unmarked). Rugose leaflet observed in progeny of a cross involving D. intortum, D. sandwicense, and D. uncinatum was recessive (Park and Rotar, 1968; Rotar and Chow, 1971; Chow and Crowder, 1973). Cytoplasmic male sterility has been found in D. sandwicense and restorer genes with a complimentary dominant action occur in D. intortum (McWhirter, 1969).

SPECIES OF KNOWN VALUE

Species used as pasture and fodder crops are discussed in the

next section whereas those reported to have value as a soil cover
for green manure, for erosion control, or as having medicinal value are listed below.

Soil Cover and Green Manures

D. adscendens, D. capitatum, D. cephalotes, D. cuneatum, D.
gangeticum, D. heterocarpon, D. ovalifolium[1], D. purpureum, D.
scorpiurus, D. tortuosum, D. triflorum, D. umbellatum.

Medicinal

D. capitatum, D. gangeticum, D. heterocarpon, D. heterophyllum, D. pulchellum, D. triflorum, D. triquetrum, D. umbellatum.

GERMPLASM COLLECTIONS

Major collections of Desmodium used for pasture and fodder
are listed in Table 3.1. Many genotypes (plants with the same
genetic composition) are common to two or more collections since
there is free exchange of material among the major research organizations. The collections listed all have substantial numbers of
D. barbatum and D. canum, although the CIAT collection with 164
and 139 individual samples (accessions) of D. barbatum and D. canum respectively contains many more entries than any other collection. It is surprising that no named varieties (cultivars) selected from these species have been released. D. canum in particular is common in native pastures in Hawaii, Colombia, and Brazil. Collections of species from which commercial cultivars have
been released contain from six (D. heterophyllum) to sixty (D. intortum) accessions. Many of these accessions were collected after
the time of cultivar release. Commercial cultivars were selected
from a much narrower genetic base than is now present in collections.

A total of eighty species is represented in the collections
listed. This is less than one quarter of the known number of
species of Desmodium and many of these are represented by only a
single entry. It is obvious that there is scope for a substantial
increase in germplasm (viable seed) collections if a reasonable
sample of the genetic resource is to be obtained.

The inclusion of large numbers of unidentified accessions in
each collection indicates the need for further taxonomic work to
document and describe the whole genus, and for taxonomists to ex-
amine and identify existing material.

In addition to the large collections, small collections, usually consisting of a set of accessions of proven value, are held
by most organizations involved in pasture and fodder plant research in tropical and sub-tropical zones. These may contain a
restricted number of species or may be for a special purpose.

---

[1]D. ovalifolium has been included in D. heterocarpon by Ohashi
(1973) but is listed here as it is widely known by this name in
southeast Asia.

Table 3.1
Major <u>Desmodium</u> Germplasm Collections Held Throughout the World

Country	Organization[1]	Number of[2] species	Number of[3] accessions	Comments
Australia	CSIRO	58	750	Species best represented are D. adscendens, D. barbatum, D. canum, D. intortum, D. sandwicense, D. tortuosum
Brazil	EMBRAPA	33	285	Almost half the collection is D. canum and D. barbatum
Brazil	IRI	34	300	Species best represented are D. barbatum, D. canum, D. discolor, D. intortum, D. sandwicense, D. tortuosum
Colombia	CIAT	20	500	Over 60% of the collection is D. barbatum and D. canum
U.S.A.	University of Florida	30	225	D. barbatum, D. canum, D. intortum, and D. uncinatum are well represented

[1] CSIRO   Commonwealth Scientific and Industrial Research Organization, Australia.
    EMBRAPA Empresa Brazileira de Pesquisa Agropecuaria.
    IRI   Instituto de Pesquisas, formerly IBEC, Inc. (International Basic Economy Corporation), New York.
    CIAT   Centro Internacional de Agricultura Tropical, Cali, Colombia.

[2] All collections contain a substantial number of unidentified accessions.

[3] Most recent data available for IRI is 1966. Data for other institutions is 1978.

CHARACTERISTICS OF THE MOST IMPORTANT <u>DESMODIUM</u> SPECIES USED AS
PASTURE AND FODDER CROPS

There is abundant and specific information on the cultivars
Greenleaf (<u>D. intortum</u>) and Silverleaf (<u>D. uncinatum</u>) but only
limited observations on other species.

It should be recognized that Greenleaf and Silverleaf were
released from a narrow germplasm base in response to the demand
for a tropical pasture legume at a time when there were no commercial cultivars available. Most of the data on their performance
have been accumulated since their release and it is only now that
we are gaining an appreciation of their rational use in pastures
and their limitations. In order to achieve this understanding
wide testing has been necessary, much of which has been at or beyond their adaptive limits. These limits could not be established
without such testing.

With the large increase in the volume of material held in
germplasm collections it is likely that superior cultivars will be
selected from <u>D. intortum</u> and D. <u>uncinatum</u>. These comments apply
to other species to a greater degree. D. <u>heterophyllum</u>, for example, is very widespread throughout Southeast Asia and the Pacific,
but is very poorly represented in collections. The commercial
cultivar, Johnstone hetero, released in 1973, was selected from
one of a set of only four accessions.

D. intortum and D. uncinatum

These two species are considered together as they have many
characteristics and requirements in common. Both are perennial
legumes with scrambling stems and have been grown in many areas of
the tropics and subtropics. There is now reasonably consistent
agreement about climatic, edaphic, and pasture management limits
within which they are likely to be useful. Our discussion of
these species is largely based on findings since 1970 as results
prior to this have been reviewed by Bryan (1969) and Skerman
(1977). Research and commercial experience has been primarily
with the cultivars Greenleaf (<u>D. intortum</u>) and Silverleaf (<u>D. un-
cinatum</u>) (Figures 3.1 and 3.2).

Climatic requirements. The species are best adapted to humid
low altitude subtropical or high altitude tropical climates. They
have not persisted in the lowland tropics. Bryan (1969) noted
that in subtropical Australia these species are only suited to
areas receiving over 1,000 mm of rain. This generalization is
supported by more recent results in Florida, Australia, Uganda,
Hawaii, and South Africa. However, established stands of both
species can be thinned in unusually dry seasons and may not subsequently recover in years of normal rainfall (e.g. Rees <u>et al.</u>,
1976). As most experiments run for less than three years it is
likely that the lower rainfall limit for long term persistence
could be above 1,000 mm and up to 1500 mm, depending on such fac-

Figure 3.1. Greenleaf desmodium sward.

tors as seasonality, reliability, and soil moisture storage.
There are differing reports on the relative drought hardiness of
the two species. Hutton (1970), Skerman (1977), and Mappledoram,
and Theron (1972) noted that Silverleaf was not as hardy as Greenleaf, but Rees et al. (1976) reported the opposite. Thomas (1975)
also found that Silverleaf persisted in a region of Malawi receiving 825 mm rainfall whereas Greenleaf did not. Probably the differences are small and of little consequence.
     Controlled environment experiments have shown that both spe-
cies have a greater tolerance of cool conditions and poorer tolerance of temperatures above 30°C than most other commercially grown
tropical legumes (Whiteman, 1968; 't Mannetje and Pritchard, 1974;
Sweeney and Hopkinson, 1975). High night temperature may be more
critical than day temperature. This is reflected by the promise
shown by these species in the subtropics of southern Japan, Hawaii, Brazil, and Australia, and in elevated areas of the tropics
in Kenya, Australia, and Thailand. Both species have failed at
low altitudes in the wet tropics of Australia and Silverleaf failed in the Solomon Islands. Cool season growth of these species in
the field is usually greater than that of other tropical legumes
(Kretschmer et al., 1973; Whiteman and Lulham, 1970). For example, regrowth commences at 12°C in Greenleaf and 14°C in Siratro.

Figure 3.2.   Silverleaf desmodium sward.

Top growth is readily frosted and young plants can be killed by frost (Jones, 1969). Silverleaf plants can survive low winter temperatures, as evidenced by their survival for three years at Palmerston North, New Zealand (Hutton, 1970) and in Kokstad, South Africa (Mappledoram and Theron, 1972) under minimum temperatures as low as -14°C.

   Soil type and fertility requirement. Greenleaf and Silverleaf are adapted to non-saline slightly acid soils of widely varying texture but where natural fertility is high or has been corrected by fertilizer. They respond strongly to phosphorus, potassium, molybdenum, and sulfur, and onset of deficiency of these nutrients can lead to rapid disappearance of the species from grass-legume swards. When depending on symbiotically fixed N they only exhibit moderate tolerance to soil conditions associated with soil acidity.

   D. intortum and D. uncinatum required twice the amount of phosphorus than did Stylosanthes humilis for maximum yield (Andrew and Robins, 1969a). White (1972) found higher relative growth rates and a corresponding lower phosphorus utilization efficiency for seedling growth in D. intortum than for Macroptilium atropurpureum or S. humilis. Similarly, in the field, high nitrogen

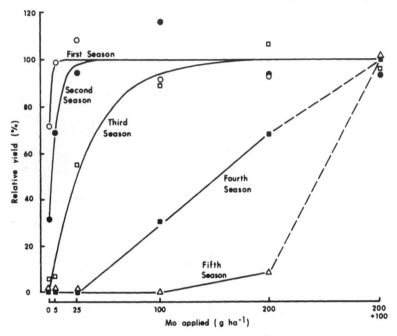

Figure 3.3. Effect of molybdenum application on relative dry matter yield of Greenleaf in different seasons. [Molyb-
denum was applied at sowing except for the maintenance treatment (200 + 100) which received 200g ha
at the beginning of each subsequent season (Johansen
et al., 1978). (-Mar. 1972, -Feb. 1972, -Nov. 1973,
-May, 1975, -Jan. 1976)]

yields per unit of applied phosphorus only occurred at high phosphorus applications (40 and 80 kg per ha P) with D. intortum
whereas high nitrogen yields were observed at lower (10 and 20 kg
P/ha) applications with S. guianensis (Blunt and Humphreys, 1970).
D. intortum has been shown to have a relatively high requirement for molybdenum (Johansen et al., 1977). On a soil where
there was no response to Mo by S. guianensis, D. intortum required
100 g Mo per ha over three years, and disappeared completely in
the absence of molybdenum application (Figures 3.3 and 3.4). Demand for potassium can increase where there is strong competition
by a grass for the nutrient (Hall, 1974). D. uncinatum was shown
to be less responsive to copper than S. guianensis (Andrew and
Thorne, 1962).
Critical plant tissue nutrient concentrations derived at the
pre-flowering stage under conditions of adequate moisture are approximately similar for both species and are 0.21-0.24 percent
for P (Andrew and Robins, 1969a), 0.7-0.8 percent for K (Andrew

Figure 3.4. Response of Greenleaf desmodium to molybdenum in a pot experiment by P.C. Kerridge. (Application rates of molybdenum were 0, 7, 35, and 875 g/ha for treatments 1 to 4, respectively.)

and Robins, 1969b) and 0.15-0.18 percent for S (Andrew, 1977). The molybdenum requirement is less than 0.02 percent, too low for routine chemical determination (Johansen, 1978b). These concentrations reflect internal requirements by the plant but they must be applied with caution in determining nutrient status as they vary with plant part and age of plant (Johansen, 1978a) and moisture availability. Andrew and Pieters (1972) have produced an illustrated booklet of foliar symptoms of nutrient deficiencies for D. intortum.

D. uncinatum and D. intortum showed poor salt tolerance when compared with other tropical legumes (Russell, 1976). In a similar manner they exhibited chloride toxicity with applications of 250 kg per ha KCl (Andre and Robins, 1969b). Further R.M. Jones (1973) has observed seedling toxicity and mortality of D. intortum sown with broadcast applications of 100 kg per ha KCl on a light textured soil.

The amounts of fertilizer required will be dependent on soil type, but, in general, superphosphate application is essential, molybdenum should be applied in deficient areas, and application of potassium at establishment should be restricted to less than 50 kg per ha K as KCl on light soils. Molybdenum may be applied at establishment in the trioxide form by incorporation with the coating used for pelleting seed, advantages being even application and lower cost (Kerridge et al., 1973). Molybdenum deficiency may not appear for two to three years on more fertile soils until available soil nitrogen reserves have been used up. Delayed response to potassium has also been observed.

Examples of fertilizer appications: On humic gley (Ultisols) and podzolic (Alfisols) soils of south-east Queensland--limiting nutrients for plant growth are P, N, Ca, K, Cu, Zn, Mo, B in descending order of importance--625 kg single superphosphate, 625 kg calcium carbonate, 125 kg potassium chloride, 8 kg copper sulfate, 8 kg zinc sulfate, 8 kg boron and 280 g sodium molybdate per hectare were recommended for establishment of Greenleaf and Silverleaf pastures (Andrew and Bryan, 1958). Subsequent annual applications of 250 kg per ha superphosphate have resulted in higher animal production than 125 kg per ha, whereas annual potassium applications of 65 kg per ha KCl have been adequate.

D. intortum pastures on krasnozem and zanthozem soils (Ultisols) deficient in phosphorus, molybdenum, and sometimes potassium, on the Atherton Tableland in north Queensland, Australia, receive at least 500 kg per ha molybdenized single superphosphate at establishment and annual maintenance applications of 250 kg per ha, with molybdenum being applied at the rate of 100 g per ha every 3 years and potassium chloride at 60-120 kg per ha per annum where deficiency occurs.

The minimum application of 250 kg per ha superphosphate gave good establishment of D. intortum and D. uncinatum oversown into grazed or burnt grasslands in Kenya (Keya and Kalangi, 1973). The application of nitrogen to legume-grass pastures usually has resulted in loss of legume in cutting experiments (Whitney, 1970); however, farmers on the Atherton Tableland, north Queensland, have made single annual applications of 50-75 kg N per ha in the spring

months to D. intortum pastures without long-term detrimental effects on the legume. In these cases the extra grass growth has
been utilized fully.

Nodulation and nitrogen fixation. Both species have specific
Rhizobium requirements, but the necessity for inoculation will depend on the presence or absence of suitable native rhizobia. The
use of an adhesive has improved survival and nodulation where inoculation is necessary, whereas pelleting with lime or other coating materials is of questionable value. Liming acid and low calcium status soils has increased yields through improved nodulation
and nitrogen fixation.

Sowing Greenleaf without inoculum can result in poor or negligible nodulation from native rhizobia and subsequent low yields
(Diatloff and Luck, 1972; Norris, 1973). Other results have shown
that although inoculation improved nodulation there was no effect
on yield after twenty weeks (Keya and Eijnatten, 1975b; Wendt,
1971). These variable effects of inoculation have been clarified
by serological studies of Date (1973) who found that in the first
year almost all nodules were from the applied strain when there
were no suitable native rhizobia present but less than 5 percent
where suitable native strains were present.

Where inoculation is necessary the work of Norris (1973)
would suggest that an adhesive, such as gum arabic or methyl cellulose, be used to apply the inoculum as it can enhance survival
of rhizobia. Rock phosphate and bauxite were shown to be suitable
coating agents for pelleting, but gave no better survival than
adhesive alone. Lime pelleting decreased survival of rhizobia on
Silverleaf after storage for one month (Norris, 1972) but had no
detrimental effect on nodulation of Greenleaf where seed was
stored one week (Diatloff and Luck, 1972).

Nodulation of Silverleaf was depressed more than that of four
other tropical legumes in the absence of lime on a low calcium
status soil of pH 5.5 (Andrew and Norris, 1961). Similarly,
Andrew (1976) found Silverleaf to be more sensitive to low pH than
several other tropical legumes. Nodulation and nitrogen fixation
were markedly reduced below pH 5.0 and pH 6.0 respectively (Andrew
and Johnson, 1976). Greenleaf was intermediate in another group
of tropical legumes but responded strongly to lime on a soil of pH
4.7 (Munns and Fox, 1977). Lime application did not affect Greenleaf yield on a soil of pH 5.6 (Olsen and Moe, 1971). Lime pelleting does not appear to have a similar role in acid soil for
Desmodium species as it does for Trifolium species. Where there
is no yield response to lime the effect of lime pelleting is transitory (Diatloff and Luck, 1972; Keya and Eijnatten, 1975b) and
where the effect does persist it is small in comparison with that
from liming the soil (Elkins et al., 1976). Silverleaf exhibited
moderate tolerance to excess manganese (Andrew and Hegarty, 1969)
and aluminum (Andrew et al., 1973) when grown with applied nitrogen but it is likely the tolerance limits would be lower for nodulation and nitrogen fixation.

Greenleaf is relatively slow to nodulate even under favorable conditions. Improvement through breeding appears remote because genetic variation in nodulating ability, shown in controlled conditions, has not been demonstrated in the field (Imrie, 1975).

Field studies have shown that nodules on Silverleaf are initially concentrated around the taproot but become dispersed with time (Whiteman, 1971). Nodule weight is markedly seasonal (Whiteman and Lulham, 1970) and is reduced by defoliation (Whiteman, 1970). In one instance nodule decay contributed 14 percent of the seasonal nitrogen release (Whiteman, 1971). Detailed pot culture studies have been made on the effects of nitrogen and shading on the growth of Greenleaf seedlings (Kitamura et al. 1976).

Nitrogen fixation in legumes is a function of nitrogen concentration and yield less uptake of mineral nitrogen. Typical nitrogen concentrations in Silverleaf and Greenleaf are 3.0 to 3.5 percent with extremes of 1.5 to 4.0 percent (Bryan, 1969). Lower N concentrations could result from inadequate soil moisture as well as inadequate nutrients - molybdenum (Johansen et al., 1977), phosphorus (Andrew and Robins, 1969a), sulfur (Andrew 1977), and low pH (Andrew, 1976). Variation in yield, however, is probably the main factor in the large range of nitrogen inputs of 89 to 382 kg N per ha quoted by Bryan (1969). One feature of Greenleaf, and presumably also Silverleaf, is that leaf litter is slow to mineralize, due to the formation of polyphenol protein complexes upon the death of plant cells (Vallis and Jones, 1973). This may account for the slower initial transfer of nitrogen to the grass in D. intortum than Macroptilium atropurpureum pastures (Vallis and Jones, 1973) but the effect is probably not important in older stands.

Seeding and seed production. Silverleaf and Greenleaf flower when daylength is less than twelve hours (Chow and Crowder, 1974). Silverleaf flowers about one month before Greenleaf (Bryan, 1969). Application of nitrogen at the flower initiation stage in a microplot experiment in southern Queensland increased seed yields of D. uncinatum by 21 to 31 percent (Gibson and Humphreys, 1973). Seed production was also increased in field experiments by high superphosphate application and a moderate application of nitrogen in favorable years (Nicholls et al., 1973), but it remained very low (5-60 kg per ha) compared to 400 kg per ha obtained in north Queensland. The climatic attributes required to make an area suitable for seed production are: annual rainfall of 800 to 2000 mm falling predominantly in summer with no more than 400 mm falling outside the four wettest months; average daily mean temperature of the coolest month above 17°C; ground frost risk low; and latitude greater than 10°C (Hopkinson and Reid, 1979). Keya and Eijnatten (1975a) found that threshing and scarification increased the germination percentage of both fresh and stored Silverleaf seed. Practical advice on seed production is given by Humphreys (1978) and Jones and Roe (1976).

111

Pests and diseases. Important pests of both Desmodium species are the weevils Amnemus quadrituberculatus and A. superciliaris in subtropical Australia. Both are prevalent on well drained
krasnozem and chocolate soils derived from basalt. The adult weevils damage leaves but the more serious damage results from larval
severing or furrowing of roots so that plants are very susceptible
to subsequent dry periods (Braithwaite and Rand, 1970). Presowing application of dieldrin or heptachlor controlled larvae and
improved persistence, but the legume population still declined
(Rand and Braithwaite 1975).

A similar problem occurs in tropical Australia with the weevil Leptopius corrugatus (Kerridge and Everett, 1975, Shaw and
Quinlan, 1978), although in this instance only Greenleaf is attacked, Silverleaf being tolerant, if not resistant to the weevil.
Many authors refer generally to 'insect attack' on both species
e.g. Gartner et al. (1974), Chow (1974). Growth of Greenleaf is
severely affected by leaf eating insects in Colombia[2] and Hawaii[3].
Greenleaf is relatively resistant and Silverleaf more susceptible
to root knot nematode (Valdez, 1975, Kretschmer et al., 1980).

The susceptibility of both species to other pests and diseases is discussed by Bryan (1969) and Skerman (1977) but there is
no firm evidence, except where indicated above, that the agronomic
use of either species is severely limited by pests or diseases.

Establishment. Greenleaf is readily established in cultivated weed-free seedbeds but poor establishment may occur in weedy
situations or when oversowing into existing grassland. Silverleaf
can be successfully oversown into grassland.

Greenleaf has very small seeds (750/gm) which may partly account for the problems in establishing this species in the face of
vigorous weed competition. Jones (1975) found that Greenleaf
yield was reduced ten-fold by the presence of dense annual grassy
weeds. Grant (1975) found that under intense weed competition,
there was a seven-fold higher death rate of Greenleaf seedlings
than of Siratro or 'fine-stem' stylo seedlings. A survey, however, showed that farmers in southeast Queensland, Australia, were
reasonably satisfied with establishment of Greenleaf in their pastures (Jones and Rees, 1973). Greenleaf was the most shade tolerant of four legumes during establishment, although absolute
yields of shaded Greenleaf were less than those of shaded Siratro
(Whiteman et al., 1974).

Tudsri and Whiteman (1977) were unable to establish Greenleaf
in an existing sward of Setaria anceps cv. Kazungula; however,
Cook and Grimes (1977) established Greenleaf in moist areas of a
burn in recently thinned Eucalyptus forest.

Experience suggests that Silverleaf with its larger seed (200

[2]Information supplied by Dr. B. Grof.
[3]Information supplied by Dr. P.P. Rotar.

per gram) is easier to establish than Greenleaf (Bryan, 1969). Silverleaf seedlings were able to penetrate the root mass in compacted topsoils of undistubed grassland in Kenya (Keya, 1976). Studies on establishing Silverleaf in Malawi have revealed a range of successful approaches, namely, (i) undersown with a maize crop (ii) sown prior to the grass, and (iii) sown alternately in strips or blocks (Thomas, 1975).

Keya et al. (1972b) compared several techniques of sowing and oversowing Silverleaf into undisturbed Hyparrhenia grassland and concluded that the cheapest and quickest method was to oversow after burning or grazing undisturbed grassland. Superphosphate application was essential, 250 kg per ha being recommended as an economic rate (Keya and Kalangi, 1973).

The optimum time for sowing Desmodium will obviously vary with each site depending on such factors as rainfall and temperature regimes and site accessibility. Greenleaf was the most susceptible of a range of tropical legumes tested for their tolerance of high temperatures during germination (Gomes and Kretschmer, 1978). In subtropical Australia different studies have suggested sowing before the onset of the growing season (Kemp, 1976), at the onset (Whiteman and Lulham, 1970), or during the rainy season (Bryan, 1969). The usual recommended sowing rates are 1-5 kg per ha for Silverleaf and 1-3 kg per ha for Greenleaf. Using higher rates confers an advantage in the establishment year, but not necessarily in subsequent years in either cultivated seedbeds (Olsen and Tiharuhondi, 1972; Middleton, 1970; Jones, 1975) or oversown undisturbed grassland (Keya and Kalangi, 1973).

Vegetative establishment of both species from rooted splits has been successful by either hand (Keya et al., 1972a) or machine planting (Younge et al., 1964), but would only have application in small scale farming (Keya et al., 1972a).

Yield and persistence under cutting. Greenleaf has been very productive under cutting, but increased frequency of cutting and lower cutting height can reduce yields.

High yields of Greenleaf have been recorded in many short term cutting experiments in pure swards and legume/grass mixtures (Bryan, 1969; Skerman, 1977). Locations where total yields of mixed swards with about 50 percent legume were above 10,000 kg per ha per year include Florida (Kretschmer et al., 1973), Uganda (Olsen 1973), Australia (Miller and Van der List, 1977), and Hawaii (Whitney, 1970). Greenleaf yields in pure swards are frequently above 6,000 kg per ha year (Jones, 1973; Stobbs and Imrie, 1976; and Clatworthy, 1975a). Very high yields can be obtained: Riveros and Wilson (1970) harvested 17,000 kg per ha from a pure Greenleaf sward and Whitney et al. (1967) 29,900 kg per ha.

Yields of 3,500 and 5,300 kg per ha per year from pure Silverleaf swards were measured at two sites in Rhodesia where corresponding Greenleaf yields were 6,570 and 6,430 (Clatworthy, 1975a). In another experiment both legumes gave similar yields of about 6,500 kg per ha (Clatworthy, 1977). Keya (1974) recorded

Figure 3.5. Hereford steers grazing a Greenleaf desmodium/Nandi
setaria pasture at Samford, Queensland, Australia.

yields of 3-5,000 kg per ha from both Greenleaf and Silverleaf in
4 year old swards at two sites in Kenya, where total yields were
9-12,000 kg per ha.

Where cutting experiments have been continued for three
years, yields have sometimes markedly declined by the second year
(Funes and Yepes, 1974; Mappledoram and Theron, 1972) or the third
year (Thomas, 1976a) although in other cases high yields have been
maintained (Jones, 1973). In two long term cutting experiments
both desmodiums persisted for six years or more, although there
was a decline in legume yield (Table 3.2).

Results from experiments in which different cutting heights
and frequencies were imposed on Greenleaf swards are summarized in
Table 3.3. Yield was depressed by cutting intervals of 3-5 weeks
compared with 8-12 weeks. Cutting height had much less effect,
though R.J. Jones (1973) found that low cutting heights (3-5 cm)
depressed legume yield under frequent cutting (4-5 weeks), but had
less effect with 8-10 week cutting, and slightly increased legume
yield with the longest cutting interval. Weed invasion into pure
swards was greater with more frequent cutting and always decreased
with increasing cutting height. Olsen (1973) also found that
total yield and Greenleaf yield increased as the cutting interval
increased from 3 to 9 weeks, with 3-weekly cutting having a more
severe effect on Greenleaf at 8 cm than at 13 cm cutting height.

TABLE 3.2
Dry Matter Yield (t/ha^{-1}) of <u>Desmodium</u> and Associated Grass in Long Term Cutting Experiments in Australia and Kenya

Year	Australia[†]		Kenya[††]		Kenya[††]	
	Greenleaf	Setaria	Greenleaf	Native grass	Silverleaf	Native grass
1967	2.4	8.8	0.3	0.8	0.3	0.9
1968	9.5	7.4	3.1	3.4	2.6	4.0
1969	7.1	5.9	3.0	4.8	2.8	4.8
1970	nm	nm	2.6	4.3	2.1	4.4
1971	5.4	4.0	2.3	3.4	2.1	3.5
1972	5.1	4.2	1.4	3.3	1.6	3.5
1973	4.7	4.2				
1974	5.2	4.0				

(nm - not measured but cut to schedule; † irrigated greenleaf/Nandi setaria swards (Wilson, 1976); †† oversown dryland <u>Hyparrhenia</u> grassland cut every eight weeks (Keya, 1976).

TABLE 3.3
Yields (L/ha⁻¹) of Pure Greenleaf Swards and Greenleaf/Grass
Swards under Different Cutting Heights and Frequencies

Reference	Cutting Interval (wk)	Cutting Height (cm)	Yield		
			Legume	Other sp.	Total
Legume/sown grass					
Whitney (1970)[1]	5	5	2.2	2.9	5.1
	5	13	4.3	3.5	7.8
	10	5	6.7	7.0	13.7
	10	13	8.4	6.8	15.2
Riveros & Wilson (1970)[2]	3	7.5	7.1	10.5	17.6
	3	15.0	6.5	10.6	17.1
	5	7.5	7.0	8.8	15.8
	5	15.0	8.8	8.0	16.8
Murphy et al. (1977)[3]	3	5	2.9	4.6	7.5
	3	13	2.2	4.2	6.4
	6	5	4.5	5.4	9.9
	6	13	4.0	4.9	8.9
Legume swards					
R.J. Jones (1973)[4]	4	3.8	3.3	4.3	7.6
	4	7.5	5.0	2.4	7.4
	4	15.0	4.6	1.4	6.0
	8	3.8	6.9	3.8	10.7
	8	7.5	6.2	2.9	9.1
	8	15.0	7.1	1.4	8.5
	12	3.8	9.3	2.6	11.9
	12	7.5	8.6	2.1	10.7
	12	15.0	7.2	1.1	8.3

([1] with kikuyu grass, Pennisetum clandestinum, yields were very
similar with pangola grass, mean of 2 year's results; [2] with Setaria sphacelata, second year's results, [3] mean of rhodes grass and
pangola grass, second year's results, [4] mean of three year's results).

Silverleaf has a similar reaction, yield being depressed by a five week defoliation interval with a similar height x interval interaction (Keya, 1976).

Although Greenleaf has yielded well under a wide range of defoliation treatments, there is a clear trend for depressed yield with more frequent and lower cutting. In contrast, Riveros and Wilson (1970), who adopted relatively close cutting intervals and higher cutting heights (Table 3.3), found little effect of these factors on yield. They observed that Greenleaf developed a 'hedge-like stand' with the height of cutting only affecting the height of this productive framework. Extending this concept to the grazed situation we would expect that survival of Greenleaf would be enhanced if a similar framework could be developed and maintained under grazing. Greenleaf swards, in common with most twining or scrambling tropical legumes, have low foliage densities whether sown alone (Stobbs and Imrie, 1976) or with a grass (Heslehurst and Wilson, 1971).

Well preserved acetic acid silage was made from Greenleaf alone, and a stable lactic acid silage was produced when 8 percent molasses was added (Catchpoole, 1970).

Yield and persistence under grazing. Greenleaf and Silverleaf pastures generally do not persist for longer than 4 to 6 years under grazing. The reasons are not always clear but appear to be a combination of heavy stocking, soil moisture stress in the growing season, nutrient deficiencies, insect attack, and poor seedling regeneration.

Yield and persistence of both species are almost always poor under heavy grazing whether rotationally grazed (Bryan and Evans, 1971a, Whiteman, 1969, Febles and Padilla, 1972) or continuously grazed (Bryan and Evans, 1973; Jones, 1974). Evidently the 'productive framework' formed under frequent cutting does not develop or is not maintained under heavy grazing. Persistence is favored by keeping the pasture between 30 and 60 cm height during the growing season (Roberts, 1979).

Under light to moderate grazing pressure both species have shown variable persistence in short term grazing experiments. In Hawaii, the mean Greenleaf percentage in a grazed pasture was 38 percent in the first year and 32 percent in the second (Younge et al., 1964). Over three years the percentage of Greenleaf in a lightly stocked pasture in Australia rose from 30 to 49 percent (Table 3.4). Silverleaf persisted well under infrequent grazing in Malawi (Thomas, 1976b), comprising 25 percent of yield after four years, however, Greenleaf failed to persist. Greenleaf also failed to persist in Rhodesia (Clatworthy, 1975a) where the percentage of Greenleaf in one paddock declined from 50 percent in one year to 2 percent, in the following, without any apparent reason. Rees et al. (1976) found that neither Greenleaf nor Silverleaf could recover following a dry summer at two sites in southeast Queensland.

TABLE 3.4
Changes in Animal Production and Legume Component in Nandi Setaria Pastures with Greenleaf Desmodium or Siratro in Southeast Queensland, Australia (Sown in 1968 and Set-Stocked at 1.1 and 3.0 Beasts ha^{-1}.)*

Sown legume	Stocking rate beasts ha	Mean liveweight gain hd^{-1}		% legume in pasture (May)[1]		
		1969-1972	1976-1978[2]	1970	1972	1979
Greenleaf	1.1	180	130	30	49	0
Siratro	1.1	160	146	51	37	15
Greenleaf	3.0[3]	70	44	23	6	0
Siratro	3.0[3]	30	37	41	9	0

*([1]legume percentage in presentation yield, [2]includes poor gains in dry 1976/77 summer, [3]stocked at 2.0 beast ha^{-1} since 1975. 1969-1972 data (Jones, 1974) and 1976-1979 data (Jones, R.M. and Jones, R.J. Unpublished).

We are only aware of two long term experiments where Desmodium pastures have been grazed for longer than five years, both being in Queensland, Australia (Figure 3.5). At Samford, Greenleaf improved with light stocking for three years, then subsequently died out (Table 3.4).

At Beerwah, Bryan and Evans (1973) reported good persistence of Greenleaf and, to a lesser extent Silverleaf, for six years at stocking rates of 1.2 and 1.7 beasts per ha; however, in a continuation of this experiment both species were virtually eliminated at 1.7 beasts per ha and the percentage of Desmodium declined at 1.2 beasts per ha. During the first seven years Greenleaf appeared to be better suited to the drier soils, but subsequent experience has been that Greenleaf has tended to die out on the drier soils during dry conditions and has not recovered, though it has persisted in wetter soils.

There is very little known about the effect of different systems of rotational grazing on Desmodium yield and persistence. Edgely and Quinlan (1973) recommended that Desmodium based dairy pastures on the Atherton Tableland in north Queensland be rested for 3 to 4 weeks between strip grazing, the pastures never being grazed below 15-22 cm from ground level. Farm pastures of Greenleaf desmodium with Setaria or Kikuyu grasses in this district have persisted for 10 years or longer under a lenient grazing system of 2-3 days grazing every 8-10 days. Under this grazing system a fixed structure of Greenleaf is maintained. Keya (1976) suggested a 5 to 8 week interval between grazings for Silverleaf based pastures in Kenya. One experienced commercial user has stressed the need for rotational grazing in subtropical Queensland (Duggan, 1979), but another recommended continuous grazing (Roberts, 1974; 1979).

On one Queensland commercial property Desmodium had persisted on moister sites whereas it had failed on similar sites on many other nearby properties. Possible reasons for such failures were insufficient maintenance superphosphate, frosting, and competition from aggressive grasses such as Setaria (Harle, 1976). On another property Greenleaf remained productive for some five years but had largely disappeared eight years after sowing (Hurwood, 1979). In a study of soils on this particular property Rayment et al. (1979) suggested failure to re-apply molybdenum as another cause of poor Greenleaf persistence. They observed that persistence was less of a problem on soils with more rapid infiltration rates and greater subsoil storage of moisture. On the Atherton Tableland, Greenleaf pastures are more vigorous and persistent in areas with high, even rainfall distribution and low incidence of frost.

Many grass species are listed as being compatible with Silverleaf and Greenleaf (Bryan, 1969; Skerman, 1977). All these results are from experiments of less than five years' duration and there is no evidence that the companion grass has a dominant effect on Desmodium persistence provided nutrient requirements are satisfied.

In addition to withstanding effects of grazing and other forms of stress, a persistent species must be able to regenerate vegetatively or by seed. Observations suggest that vegetative reproduction from rooted stems, favored by continuously moist soil, is the main mechanism of Greenleaf regeneration (Jones, 1980). There is no information on the nature and longevity of secondary roots as compared to the primary tap root. There is little information on Greenleaf regeneration from seedlings. Limited data from subtropical Australia showed that the reserves of Greenleaf seed in the top 10 cm soil beneath good legume stands were as high as 6000 seeds m   (Jones and Evans, 1977; Jones, 1980). No seed was found beneath stands from which Greenleaf had been eliminated by overgrazing. Seedling survival in established pastures has not been studied but studies on initial establishment, as noted earlier, have shown that seedlings are susceptible to competition. Thus, if a pasture is grazed lightly enough to allow seeding, the same pasture may severely limit survival of new seedlings unless special management procedures are adopted. There is no information dealing with regeneration in Silverleaf.

Renovation of rundown pastures in Australia by cultivation or a single cropping phase has not resulted in good regeneration of Greenleaf, whereas it does aid regeneration of Siratro (Hurwood, 1979).

We conclude from these studies and observations that overgrazing is the main factor which limits Desmodium persistence. A continuous supply of soil moisture, at least in the growing season, is required for long term persistence. Deleterious effects of dry periods are more severe when there is overgrazing. Nutrient deficiency and root damage from weevils will interact with both these factors. Further, it is likely that persistence is limited by poor regeneration. More detailed studies are required to understand the mechanism of Desmodium persistence if poor persistence is to be overcome by either breeding or management.

Feed quality and animal production. Information on chemical compostion of both species has been tabulated by Skerman (1977). The typically low in vitro digestibility of 55-60 percent (Stobbs and Imrie, 1976) is partly explained by the presence of polyphenols (Ford, 1978). Soluble carbohydrate levels of Greenleaf are much lower than for white clover, with ranges of 0.5-3.0 percent and 2.5-8.0 percent respectively (Noble and Lowe, 1974). Quality and quantity of bulked Desmodium feed can decline markedly over winter (Lowe, 1976). There is no evidence for oestrogenic activity in Greenleaf (Little, 1976).

Milk production from Jersey cows grazing pure Greenleaf was only 7.7 kg per cow per day (Stobbs 1971). This low milk yield occurred with pure stands and has been partly attributed to the difficulty experienced by cows in obtaining sufficient feed from the loose legume sward during their grazing period (Stobbs and Thompson, 1975). Much higher milk production has been obtained on farms on the Atherton Tableland in north Queensland from mixed Greenleaf-grass pastures.

Good beef production has been recorded from Desmodium pastures where there was a strong legume component, usually greater than 30 percent. The highest weight gain reported is that of 780 kg live weight per ha in Hawaii (Younge et al., 1964). Bryan (1969) quoted Younge and Plucknett (1965) as having achieved 1410 kg LWG per ha but the original paper states that because of poor Greenleaf establishment the pasture was fertilized with nitrogen.

Some examples of weight gains per hectare are shown in Table 3.5. Corresponding weight gains per head were 160-250 kg where animals grazed year long on good Desmodium pastures. Weight gains obviously depend on factors such as stocking rate and grazing system; our aim has been to illustrate the high levels of production that can be achieved.

Most of the animal production data recorded in Table 3.5 has been from short term experiments. In longer term experiments weight gain has declined as the Desmodium content of the pastures decreased (Table 3.4, Bryan and Evans, 1971b). Good Silverleaf and Greenleaf pastures can support profitable beef production enterprises (Keya, 1976; Firth et al.; 1975, Nel and Bosman, 1979) and the meat has an acceptable flavor (Park and Minson, 1972).

Conclusions. Both Greenleaf and Silverleaf have shown promise in the wetter subtropics and in the more elevated tropics. Both species can be highly productive and have, at least in the short term, given good animal production. Their nutrient and Rhizobium requirements are relatively well known. The primary limitation appears to be that of poor persistence under grazing with reasons for this only partly understood. Other deficiencies include susceptibility to weevil attack and, in the case of Greenleaf, poor seedling vigor.

D. heterophyllum

D. heterophyllum (hetero) is a rhizomatous perennial species (Skerman, 1977) that is showing considerable promise as a pasture legume in wet tropical lowlands of northern Australia, Fiji, the Solomon Islands, the Amazon Basin, and Sri Lanka. All of these sites have an annual rainfall above 1500 mm. Hetero copes well with waterlogged sites (Gutteridge and Whiteman, 1978), but is sensitive to dry periods (Partridge, 1975).
Hetero responds to superphosphate application (Partridge, 1975). It grows well on fertile alluvial soils (Gutteridge and Whiteman, 1978), suggesting that, like Silverleaf and Greenleaf, it requires high fertility. However, experience in Malaysia showed it was more tolerant of acid soil conditions than these cultivars as it gave high nitrogen fixation on a soil of pH 4.5. In Fiji, Partridge (1975) observed that superphosphate application to natural pastures increased the D. heterophyllum component but it had lower yields (2.0 t/ha) than the oversown M. atropurpureum (8.0 t/ha). D. heterophyllum has been shown to utilize rock phosphate as effectively as S. guinanesis and more effectively than C. pubescens. This is in contrast to D. uncinatum which makes poor use of rock phosphate (Bryan and Andrew, 1971).
During 6-8 week cutting intervals hetero has yielded 1,500-3,500 kg of dry matter per ha in adapted sites (Partridge, 1975; Gutteridge and Whiteman, 1978) but yields could well be increased by more frequent and closer cutting. Under high stocking rates where Siratro has failed in grazing trials in Fiji, hetero has thickened up and formed good swards (Partridge, 1980). Hetero's ability to thrive under heavy grazing (Harding and Cameron, 1972) is such that in northern Australia it is the only legume that can combine with well grazed signal grass Brachiaria decumbens (Loch, 1977). In Malaysia, hetero persisted well in heavily grazed small plots but was slow to establish in larger areas, possibly because

TABLE 3.5
Animal Live-Weight Gains Reported from Cattle Grazing Pastures
Containing D. intortum and D. uncinatum

Country	LWG ha^{-1}	Associated Grass	Duration of Trial (yrs)	Comments
D. intortum				
1 Hawaii	778[1]	pangola	2	adequately fertilized treatments only
2 Australia	586		2	D. intortum irrigated
3 South Africa	544	rhodes grass	5	no pasture data given
4 Australia	525	pangola	5	D. intortum declined
5 Uganda	515	Panicum maximum	-	no pasture data given
6 Australia	282	pangola	6	set stocked (1.25 beasts ha^{-1})
7 Australia	245	Nandi setaria	3	1.7 beasts ha^{-1}, see also Table 3.4
8 Rhodesia	236		1	
D. uncinatum				
9 Kenya	488	oversown Hyparrhenia	1	8.5 months grazing period
Kenya	691	rhodes grass	1	8.5 months grazing period

[1]This experiment is apparently the same one as was reported giving
660 kg 'beef' ha^{-1} by Moomaw and Takahashi, 1962.

References

1 Younge et al., 1964
2 Miller and van der List, 1977
3 Nel and Bosman, 1979
4 Bryan and Evans, 1971b
6 Evans and Bryan, 1973
7 Jones, 1974
8 Clatworthy, 1975a
9 Keya, 1976

of absence of effective rhizobia. It often forms a prominent component, together with D. triflorum, of closely grazed natural grasslands such as roadsides and under coconut plantations in S.E. Asia and the Pacific.

Animals grazing hetero pastures at 4.3 beasts per ha gained 860 kg liveweight per ha in northern Australia (Harding and Cameron, 1972) and have produced over 500 kg per ha in Fiji (Partridge, 1980).

A major limitation to the wide use of hetero is its low seed recovery and, hence, high seed cost. However, the plant can be established vegetatively by mechanized equipment (Harding and Cameron, 1972), and usually spreads rapidly. The only disease problem noted on hetero so far has been legume little leaf, but it was of no agronomic consequence (Harding and Cameron, 1972). One cultivar, cv. Johnstone, released in Australia in 1973, is strain specific for Rhizobium but nodulates readily (Anon., 1973) with the correct strain. Work at CIAT, Colombia, has also demonstrated considerable strain specificity and responses to rhizobia inoculation (Halliday, 1979).

Although there is relatively little knowledge about factors such as the nutritional requirements of hetero and its yield under different management regimes, there is optimism about its agronomic future in those areas to which it is adapted. Good animal production can certainly be obtained from good stands and current experience suggests that with adequate grazing pressure and fertilization there is no serious difficulty in maintaining productive hetero pastures.

D. canum

D. canum (Kaimi clover) has been described in detail by Skerman (1977) and also by Younge et al. (1964) who reported very encouraging plant and animal production from this species in Hawaii. It is a perennial plant, which becomes prostrate under grazing, and is compatible with grasses such as pangola grass (Digitaria decumbens), and which requires a rainfall in excess of 1250 mm. Surprisingly, there are very few subsequent reports on this species. Yields from D. canum were lower than from D. intortum in Hawaii, averaging two and seven tons per ha, respectively, but this would be expected under the eight to thirteen week cutting frequency imposed (Whitney and Green, 1969). It invades improved pastures in south Florida and can comprise 15-25 percent of the botanical composition by weight[4]. Kretschmer (1971) refers to native stands being utilized in Brazil. Collections from Brazil and Colombia are being evaluated in those countries (Anon., 1976). D. canum has shown some promise in Rhodesia (Clatworthy, 1975b) but is unsuited to the wet tropics of Australia (Teitzel et al., 1974). D. canum is susceptible to a virus disease in Hawaii (Edwardson et al., 1970).

---

[4] Information supplied by Dr. A.E. Kretschmer.

D. canum responded to lime, molybdenum, zinc, phosphorus and potassium in Hawaii (Younge et al., 1964). When adequately fertilized it contributed over 10 percent of dry matter yield in a seven year old pasture, yet persisted when 'underfertilized and overgrazed'. Animals grazing D. canum-grass pastures produced almost 680 kg LWG per ha, averaged over four years (Younge et al., 1964).

We cannot be sure whether Kaimi clover has been tried more widely and failures have not been reported. In trials in southeast Queensland extensive variation was found in both morphological and agronomic characters but no high yielding types were identified[5].

D. heterocarpon and D. ovalifolium

A cultivar of D. heterocarpon, referred to as 'Florida' carpon, has recently been released as a pasture legume in southern Florida, U.S.A. (Kretschmer et al., 1976, 1979). It requires reasonably high rainfall but cannot withstand waterlogging. Florida carpon withstands heavy grazing and produces abundant seed but is susceptible to root knot nematode; other accessions have nematode resistance (Kretschmer et al., 1980). It responds well to phosphorus application, and on an infertile sandy soil, gave a three-fold response to 2200 kg lime per ha when given adequate superphosphate but was depressed with 3400 kg lime per ha (Snyder et al., 1978).

D. ovalifolium has been included in D. heterocarpon by Ohashi (1973) but workers with an agronomic knowledge of these species consider that the material distributed under each name is agronomically distinct.

D. ovalifolium is included in cover crop mixtures in Malaysia and is now used in pasture mixtures. It has been slow to establish but with its stoloniferous habit forms a good ground cover. It makes better growth under heavy shade and frequent defoliation than most other tropical legumes (Lu et al., 1969) and so has potential as a pasture and cover plant in plantation crops. It was considered to have low potential as a pasture legume in Rhodesia (Clatworthy, 1975b), and in the wet tropics of Australia (Teitzel et al., 1974). It is currently being evaluated at CIAT, Colombia, (Figure 3.6) where it produced 7,000 kg per ha dry matter over eight months (Anon., 1978). It has specific rhizobia requirements and responds to inoculation (Halliday, 1979), and apparently has good drought tolerance and no specific problems caused by diseases or insects (Schultze-Kraft and Giacometti, 1979).

The recent interest in D. ovalifolium is because of its good adaptation to the highly weathered acid soils of the tropics and its persistence under grazing. It remains to be seen if good animal production can be achieved from D. ovalifolium pastures.

---

[5]Information supplied by Mr. R.J. Williams.

Figure 3.6.   Desmodium ovalifolium/Brachiaria spp. mixture.   (Note
the Stylosanthes capitata seedlings emerging from
faeces brought in by grazing animals.)

D. sandwicense

This species is subtropical in its temperature requirements
which are similar to Greenleaf and Silverleaf (Whiteman 1968,
Sweeney and Hopkinson, 1975).   It has sometimes been confused with
Silverleaf.   It is a more erect and drought resistant perennial
than Greenleaf, showing promise in areas too dry for Greenleaf and
Silverleaf (Cameron and Mullaley, 1969; Smith, 1977) and perform-
ing better in the dry season (Naveh and Anderson, 1967).   It has a
day-neutral flowering response (Rotar et al., 1967) and flowers
throughout the year in areas where moisture availability and temperatures are adequate.   In subtropical areas it flowers from mid-
summer until winter when growth ceases.   Severe insect attack has
been noted in Singapore (Chow, 1974) and Swaziland (I'Ons, 1967).
The few reported agronomic findings are often inconsistent
but this is not unexpected considering the wide variability in
growth habit and forage production that occurs in the species.
Reasonable palatability was reported from Kenya (Bogdan and Mwakha, 1970) but low palatability from Singapore (Chow and Crowder,
1973).   It is resistant or tolerant to legume little leaf disease

(Chow and Crowder, 1973) but susceptible to another virus disease[6]. D. sandwicense is usually described as low yielding relative to Greenleaf and Silverleaf (Bogdan and Mwakha, 1970; Naveh
and Anderson, 1967), but Stobbs and Imrie (1976) measured yields
of 6300 and 7500 kg per ha per yr, similar to that of Greenleaf.
Similar yields have been measured in Florida (Kretschmer, 1970)
and Rhodesia (Clatworthy, 1975a).

OTHER SPECIES

D. adscendens. In the subtropics this species flowers late
in the growing season and seed production can be limited by
frosts. It is being tested in Colombia (Anon., 1976) and in Florida (Kretschmer et al., 1980) where it is resistant to root knot
nematode.

D. barbatum. A perennial species which is grazed where it
occurs naturally in South America (Kretschmer 1971). Selections
are under test in Colombia (Anon., 1976). D. barbatum, which can
grow on acid infertile soils, includes ecotypes with a wide range
of growth habits, foliage density, and productivity (Schultze-
Kraft and Giacometti, 1979).

D. cajanifolium. An erect heavy seeding plant adapted to
acid infertile soils with potential as a browse plant (Schultze-
Kraft and Giacometti, 1979).

D. distortum. Has shown some potential as a pasture plant in
Rhodesia (Clatworthy, 1975b), Fiji (Roberts, 1970), and Australia
(Cameron and Mullaley, 1969).

D. leiocarpum. Has shown promise in Ghana, having greater
tolerance to dry periods than Centrosema pubescens (Tettah, 1972).

D. polyneurum. Can be a useful source of dry season feed in
arid areas of northern Australia where it grows near waterholes,
but digestibility is very low (Falvey, 1976).

D. rensonii. A fodder tree which outyielded and established
more rapidly in Tanzania than leucaena (Leucaena leucocephala)
(Naveh, 1971), but had only moderate potential in Rhodesia (Clatworthy, 1975b).

D. tortuosum. Commonly known as Florida beggarweed, it is an
important weed of peanuts in southern U.S.A. and has shown some
potential as a pasture plant in Rhodesia (Clatworthy, 1975b) and
Australia (Cameron and Mullaley, 1969) but not in Hawaii where it
is highly susceptible to virus[7]. D. tortuosum is very responsive
to phosphorus and potassium (Hoveland et al. 1976).

D. triflorum. This species is widespread in the tropics and
subtropics (Williams and Yunus, 1975). In Fiji it is even more
resistant to heavy grazing than D. heterophyllum but animal production suffered at such grazing pressures (Partridge, 1980). It
has also invaded fertilized, heavily grazed pastures in subtropical Australia but appears to have little impact on pasture or animal performance.

---

[6]Information supplied by Dr. P.P. Rotar.
[7]Information supplied by Dr. P.P. Rotar.

PLANT IMPROVEMENT

Despite the many Desmodium species that have been used in agriculture, and the variety of uses such as pasture component, fodder crop, and ground cover for erosion control and soil improvement, there have been few attempts at plant improvement through breeding. This does not mean that there has been no conscious selection within available material to obtain the most productive strains.

The first step in any improvement program is the acquisition of a set of genotypes through field plant collection and introduction from existing collections. The main Desmodium cultivars in use throughout the world have been selected from germplasm collections following testing programs. Silverleaf (D. uncinatum), Greenleaf (D. intortum), and Johnstone hetero (D. heterophyllum) were selected in Australia. Silverleaf originated in Brazil, Greenleaf is a mixture of three accessions from El Salvador, Guatemala (via Hawaii), and the Philippines (original source unknown); whereas Johnstone hetero came from New Guinea. Florida carpon (D. heterocarpon) was released in U.S.A. from an introduction from India. Each of these releases was selected from a small set of accessions. For example, Silverleaf was the only D. uncinatum accession in the CSIRO collection at the time of its release, and Greenleaf was a combination of three accessions from a collection of twelve. There are possibilities of obtaining better cultivars simply by sampling a larger set of accessions.

Several researchers have studied hybridization, both intra- and inter-specific, and basic information on crossing techniques and limits to hybridization has been published (Rotar et al., 1967, Chow and Crowder, 1973) for those species having the greatest known potential as pasture plants. McWhirter (1969) discovered cytoplasmic male sterility in D. sandwicense and restorer genes in D. intortum and proposed a system for breeding hybrids involving the commercial use of F2 seed. Although there is scope for breeding, no cultivars have been released from breeding programs and currently plant breeding activity is minimal.

For plant breeding to be potentially useful one must identify a deficient character, demonstrate a heritable basis and a potential response to selection, and have available an efficient and effective selection technique.

Some deficiencies which have been identified in Greenleaf desmodium, the most widely grown commercial cultivar, are as follows: small seed size with a consequent small seedling leading to problems of establishment and possibly regeneration, late flowering in the subtropics with a consequent reduction in seed production, and susceptibility to legume little leaf (LLL) disease (Imrie, 1972a; 1972b; 1973). For each of these characters improvement through breeding is possible. Greenleaf will hybridize readily with D. sandwicense which has seeds approximately three times larger (4.5 g/1000 vs. 1.5 g/1000), and the measurement of seed size in hybrid progeny is a simple procedure. Likewise, heritable variation for LLL disease tolerance occurs within D. intortum making breeding for this character feasible.

Similarly, heritable variation for nodulating ability and nitrogen fixing capacity has been found both in D. intortum and D.
sandwicense inoculated with an effective strain of Rhizobium (Hutton and Coote, 1972). Selection for improved seedling nodulation
was done on plants grown on agar, but there was a low correlation
with field performance (Imrie, 1975). Further research could
establish effective techniques for improving seedling nodulation
which could aid plant establishment. Nitrogen fixing capacity of
a mature plant stand appears to be correlated with plant yield
provided the plants are effectively nodulated and supplied with
essential nutrients. Selection for nitrogen fixation could be
most efficiently achieved by selection for plant yield.

The above characters are largely independent of environment
whereas deficiencies in other characters may be localised. For
example, Greenleaf is susceptible to many insect pests, and some
of these have a limited distribution. Resistance is unknown and
this could be due to insufficient screening or a lack of resistance mechanisms in the species. In either case breeding may not
be feasible. If there is no heritable variation then there is no
basis for breeding. Alternatively, due to the localised nature of
a problem, breeding may not be justified on economic grounds.

Knowledge of the extent of heritable variation in desmodiums
is limited, even in those species which are widely used in agriculture. In most characters of economic importance, for which
data are available, there is variation and the opportunity for improvement through breeding. An increase in use of desmodiums as
pasture plants or cover crops will undoubtedly lead to increased
agricultural production in the tropics and sub-tropics, but their
potential value would be magnified by a substantial plant breeding
input aimed at increasing and maintaining productivity. Whether
this is achieved by the removal of specific limits such as insect
susceptibility, or an improvement in quantitative characters like
yield or nitrogen fixing capacity is not important. Both are
feasible and can lead to tangible benefits.

CONCLUSIONS

Desmodium is a large genus regardless of whether the broad
taxonomic interpretation of Ohashi (1973), or a narrow view, is
taken. The genus is widespread throughout the tropical and subtropical regions of the world. There is a need for further taxonomic studies, particularly of the American species, as doubts
exist on the classification of even the most widely used cultivar
(Greenleaf) which should possibly be D. aparines rather than D.
intortum. Any confusion in taxonomy will inevitably lead to
confusion in reconciling agronomic information from different
sources.

Any taxonomic review must be preceded by further plant collection. Herbarium specimens were not surveyed for our review and
germplasm collections contain only a small portion of known species, and many of these are poorly represented. Any future taxonomic review should be based on the widest possible sample of

material and, where possible, take into account the agricultural potential of the genus.

It would aid agricultural evaluation if all published results included precise indentification of the material under study. Where appropriate, an accession number of one of the major collections would be preferred. Confusion could then be avoided if subsequent studies indicated that an accession had been incorrectly identified. Our experiences suggest that this is a real problem. Also, the possible application of published data to other areas could be aided by the provision of climatic, soil, and nutrition data.

It is apparent from this review that Desmodium is a genus that warrants further examination. The bulk of the available information on agricultural utilization applies to D. intortum and D. uncinatum, and indeed to two cultivars which have been selected from these species. Both species are highly productive where adapted, but available data indicate that climatic and soil nutritional limits within which they are adapted are relatively narrow. This observation probably applies to other species such as D. heterophyllum, D. heterocarpon and D. canum, and may account for some of the inconsistencies in reports in the scientific literature.

It is not possible to make broad generalizations about the genus. Extensive studies have been made on few species, and those studies which have been done indicate differences in their adaptation and performance in pastures. Successful use in pastures requires an understanding of climatic and edaphic limitations of each species and management to ensure that conditions are favorable for establishment and growth. Silverleaf can be successfully oversown in grassland but Greenleaf requires a prepared seed bed for good establishment. The difficulty of harvesting seed from Johnstone hetero means that it needs to be vegetatively propagated.

Greenleaf, Silverleaf, and Johnstone hetero, compared with the other tropical legumes, require relatively high soil nutrient levels for good establishment and maximum production. High nitrogen availability, which gives a competitive advantage to associated grasses, could lead to loss of the legume unless careful pasture management is practiced. Greenleaf and Silverleaf also have a low tolerance to salinity.

The use of Rhizobium inoculum to ensure good nodulation is required when suitable native strains are not present in the soil. This is particularly important with Johnstone hetero which is strain specific. Nodulation can be slow in Greenleaf but established stands having adequate nutrition can fix up to 380 kg N per ha per annum.

Well grown Desmodium pastures can support high levels of beef and milk production. Care must be taken to avoid overgrazing of Greenleaf and Silverleaf but Johnstone hetero, Florida carpon, and D. canum can withstand heavy grazing.

Considering the promising results that have been obtained for forage production and nitrogen fixation from a few widely tested species, and given the magnitude of the potentially useful portion of the available resource, there is ample scope for the future

study of the agricultural value of Desmodium   The authors hope
that this review will provide a useful background and some guidelines for future work.

REFERENCES

Andrew, C.S. 1976. Effect of calcium, pH and nitrogen on the
    growth and chemical composition of some tropical and temper-
    ate pasture legumes. 1. Nodulation and growth. Australian
    Journal of Agricultural Research 27: 611-623.
Andrew, C.S. 1977. The effect of sulphur on the growth, sulphur
    and nitrogen concentrations, and critical sulphur concentra-
    tions of some tropical and temperate pasture legumes. Aus-
    Australian Journal of Agricultural Research 28: 807-820.
Andrew, C.S. and Bryan, W.W. 1958. Pasture studies on the coastal
    lowlands of subtropical Queensland. III. The nutrient re-
    quirements and potentialities of Desmodium uncinatum and
    white clover on a lateritic podzolic soil. Australian Jour-
    nal of Agricultural Research 9: 267-285.
Andrew, C.S. and Hegarty, M.P. 1969. Comparative responses to
    manganese excess of eight tropical and four temperate pasture
    legume species. Australian Journal of Agricultural Research
    20: 687-696.
Andrew, C.S. and Johnson, A.D. 1976. Effect of calcium, pH and
    nitrogen on the growth and chemical composition of some trop-
    ical and temperate pasture legumes. II. Chemical composition
    (calcium, nitrogen, potassium, magnesium, sodium and phospho-
    rus). Australian Journal of Agricultural Research 27: 625-
    636.
Andrew, C.S., Johnson, A.D., and Sandland, R.L. 1973. Effect of
    aluminium on the growth and chemical composition of some
    tropical and temperate pasture legumes. Australian Journal
    of Agricultural Research 24: 325-339.
Andrew, C.S. and Norris, D.O. 1961. Comparative responses to cal-
    cium of five tropical and four temperate pasture legume spe-
    cies. Australian Journal of Agricultural Research 12: 40-55.
Andrew, C.S. and Pieters, W.H.J. 1972. Foliar symptoms of mineral
    disorders in Desmodium intortum. Division of Tropical Pas-
    tures, CSIRO, Australia. Technical Paper No. 10.
Andrew, C.S. and Robins, M.F. 1969a. The effect of phosphorus on
    the growth and chemical composition of some tropical pasture
    legumes. 1. Growth and critical percentages of phosphorus.
    Australian Journal of Agricultural Research 20: 665-674.
Andrew, C.S. and Robins, M.F. 1969b. The effect of potassium on
    the growth and chemical composition of some tropical and tem-
    perate pasture legumes. 1. Growth and critical percentages
    of potassium. Australian Journal of Agricultural Research
    20: 999-1007.
Andrew, C.S. and Thorne, P.M. 1962. Comparative responses to cop-
    per of some tropical and temperate pasture legumes. Austra-
    lian Journal of Agricultural Research 13: 821-835.
Anon. 1973. Register of Australian Herbage Plant Cultivars. Des-

modium heterophyllum DC. (Hetero) cv. Johnstone. Journal of the Australian Institute of Agricultural Science 39: 147-148.

Anon. 1976. Annual report, 1976. CIAT, Colombia p. C-11.

Anon. 1978. Annual report, 1978. CIAT, Colombia pp. A15-40.

Blunt, C.G. and Humphreys, L.R. 1970. Phosphate response of mixed swards at Mt. Cotton, south-eastern Queensland. Australian Journal of Experimental Agriculture and Animal Husbandry 10: 431-441.

Bogdan, A.V. and Mwakha, E. 1970. Observations on some grass/legume mixtures under grazing. East African Agriculture and Forestry Journal 36: 35-38.

Braithwaite, B.M. and Rand, J.R. 1970. The pest status of Amnemus species in tropical legume pastures in north coastal New South Wales. Proceedings of the XI International Grassland Congress, Surfers' Paradise pp. 676-681.

Bryan, W.W. 1969. Desmodium intortum and Desmodium uncinatum. Herbage Abstracts 39: 183-191.

Bryan, W.W. and Andrew, C.S. 1971. The value of Nauru rock phosphate as a source of phosphorus for some tropical pasture legumes. Australian Journal of Experimental Agriculture and Animal Husbandry 11: 532-535.

Bryan W.W. and Evans, T.R. 1971a. Grazing trials on the Wallum of south-eastern Queensland. 3. A nursery grazed by sheep. Australian Journal of Experimental Agriculture and Animal Husbandry 11: 633-639.

Bryan, W.W. and Evans, T.R. 1971b. A comparison of beef production from nitrogen fertilized pangola grass and from a pangola grass-legume pasture. Tropical Grasslands 5: 89-98.

Bryan, W.W. and Evans, T.R. 1973. Effects of soils, fertilizers and stocking rates on pastures and beef production on the Wallum of South East Queensland. 1. Botanical composition and chemical effects on plants and soils. Australian Journal of Experimental Agriculture and Animal Husbandry 13: 516-529.

Catchpoole, V.R. 1970. Laboratory ensilage of three tropical pasture legumes - Phaseolus atropurpureus, Desmodium intortum, and Lotononis bainesii. Australian Journal of Experimental Agriculture and Animal Husbandry 10: 568-576.

Cameron, D.G. and Mullaley, J.D. 1969. The preliminary evaluation of leguminous plants for pasture and forage in subcoastal central Queensland - 1962-69. Plant Introduction Review 6(2): 29-54.

Chow, K.H. 1974. Morphology and ecology of some introduced herbaceous legumes. The Gardens Bulletin, Singapore 27: 85-94.

Chow, K.H. and Crowder, L.V. 1972. Hybridization of Desmodium canum (Gmel.) Schin. and Thell. and D. uncinatum (Jacq.) DC. Crop Science 12: 784-785.

Chow, K.H. and Crowder, L.V. 1973. Hybridization of Desmodium species. Euphytica 22: 399-404.

Chow, K.H. and Crowder, L.V. 1974. Flowering behavior and seed development in four Desmodium species. Agronomy Journal 66: 236-238.

Clatworthy, J.N. 1975a. An assessment of animal production from a dryland pasture of Desmodium intortum. Department of Research and Specialist Services, Division of Livestock and Pastures, Rhodesia. Annual Report 1974/75 p. 137, 153.

Clatworthy, J.N. 1975b. Introduction and preliminary screening of pasture legumes at Marandellas, Rhodesia, 1967-73. Grassland Society of Southern Africa Proceedings 10: 57-63.

Clatworthy, J.N. 1977. A comparison of Desmodium spp. Department of Research and Specialist Services, Division of Livestock and Pastures, Rhodesia. 1976/77 Annual Report p. 142.

Cook, B.G. and Grimes, R.F. 1977. Multiple land use of open forest in south-eastern Queensland for timber and improved pasture: establishment and early growth. Tropical Grasslands 11: 239-245.

Date, R.A. 1973. Rhizobium ecology. Division of Tropical Agronomy, CSIRO, Australia. 1972-73 Annual Report pp. 57-58.

Diatloff, A. and Luck, P.E. 1972. The effects of the interactions between seed inoculation, pelleting and fertilizer on growth and nodulation of desmodium and glycine on two soils in south-east Queensland. Tropical Grasslands 6: 33-38.

Duggan, W. 1979. Introduction to Tinana development. Tropical Grasslands 13: 179-180.

Edgley, W.H.R. and Quinlan, T.J. 1973. Managing tropical pastures on the Atherton Tableland. Queensland Agricultural Journal 99: 293-297.

Edwardson, J.R., Purcifull, D.E., Zettler, F.W., Christie, R.G., and Christie, S.R. 1970. A virus isolated from Desmodium canum: Characterization and electron microscopy. Plant Disease Reporter 54: 161-164.

Elkins, D.M., Olsen, F.J., and Gower, E. 1976. Effects of lime and lime-pelleted seed on legume establishment and growth in south Brazil. Experimental Agriculture 12: 201-206.

Evans, T.R. and Bryan, W.W. 1973. Effects of soils, fertilizers and stocking rates on pastures and beef production on the Wallum of south-eastern Queensland. 2. Liveweight change and beef production. Australian Journal of Experimental Agriculture and Animal Husbandry 13: 530-536.

Falvey, L. 1976. Desmodium polyneurum - a useful native legume of Northern Australia. Journal of the Australian Institute of Agricultural Science 42: 130.

Febles, G. and Padilla, C. 1972. Effect of grazing on associations of gramineae and tropical legumes. Revista Cubana de Ciencia Agricola 6: 385-390.

Firth, J.A., Evans, T.R., and Bryan, W.W. 1975. Effects of soils, fertilizers and stocking rates on pastures and beef production on the Wallum of south-east Queensland. 4. Budgetary appraisals of fertilizer and stocking rates. Australian Journal of Experimental Agriculture and Animal Husbandry 15: 531-540.

Ford, C.W. 1978. In vitro digestibility and chemical composition of three tropical pasture legumes, Desmodium intortum cultivar Greenleaf, Desmodium tortuosum and Macroptilium atropurpureum cv. Siratro. Australian Journal of Agricultural Research 29: 963-974.

Funes, F. and Yepes, S. 1974. Pasture Introduction in Cuba. Proceedings of the XII International Grasslands Congress, Moscow. 3 (part 2): 759-773.

Gartner, J.A., Ferguson, J.E., Walker, R.W., and Goward, E.A. 1974. Evaluating perennial grass/legume swards on the Atherton Tableland in North Queensland. Queensland Journal of Agricultural and Animal Sciences 31: 1-17.

Gibson, T.A. and Humphreys, L.R. 1973. The influence of nitrogen nutrition of Desmodium uncinatum on seed production. Australian Journal of Agricultural Research 24: 667-676.

Gomes, D.T. and Kretschmer, A.E. 1978. Effect of three temperature regimes on tropical legume seed germination. Proceedings of the Soil and Crop Science Society of Florida 37: 61-63.

Grant, P.J. 1975. Pasture legume establishment on arable land. Department of Research and Specialist Services, Division of Livestock and Pastures, Rhodesia. 1974/75 Annual Report pp. 140-143.

Gutteridge, R.C. and Whiteman, P.C. 1978. Pasture species evaluation in the Solomon Islands. Tropical Grasslands 12: 113-126.

Hacker, J.B. 1968. Failure of chromosome pairing in a probable Desmodium intortum x D. sandwicense hybrid. Australian Journal of Botany 16: 545-550.

Hall, R.L. 1974. Analysis of the nature of interference between plants of different species. II. Nutrient relations in a Nandi Setaria and Greenleaf Desmodium association with particular reference to potassium. Australian Journal of Agricultural Research 25: 749-756.

Halliday, J. 1979. Field responses by tropical forage legumes to inoculation with Rhizobium. In "Pasture Production in Acid Soils of the Tropics" (eds.) P.A. Sanchez and L.E. Tergas CIAT, Cali, Colombia pp. 123-137.

Harding, W.A.T. and Cameron, D.G. 1972. New pasture legumes for the wet tropics. Queensland Agricultural Journal 98: 394-406.

Harle, J.G. 1976. Farm Visit - Messrs J.K. and J. Maher, Pinbarren. Tropical Grasslands 10: 226-228.

Heslehurst, M.R. and Wilson, G.L. 1971. Studies on the productivity of tropical pasture plants. III. Stand structure, light penetration and photosynthesis in field swards of setaria and Greenleaf desmodium. Australian Journal of Agricultural Research 22: 865-878.

Hopkinson, J.M. and Reid, R. 1979. Significance of climate in tropical pasture/legume seed production. In "Pasture Production in Acid Soils of the Tropics" (eds.) P.A. Sanchez and L.E. Tergas CIAT, Cali, Colombia pp. 343-360.

Hoveland, C.S., Buchanan, G.A., and Harris, M.C. 1976. Response of weeds to soil phosphorus and potassium. Weed Science 24: 194-201.

Humphreys, L.R. 1978. Tropical pasture seed production. Second Edition. FAO: Rome.

Hurwood, R. 1979. Deterioration and renovation of pastures at

"Bungawatta". Tropical Grasslands 13: 181-182.

Hutchinson, J. 1964. The genera of flowering plants (Angiospermae). Dicotyledones Volume 1 (Clarendon Press: Oxford).

Hutton, E.M. 1970. Tropical Pastures. In Advances in Agronomy
22: 2-74.

Hutton, E.M. and Coote, J.N. 1972. Genetic variation in nodulating ability in Greenleaf desmodium. Journal of the Australian Institute of Agricultural Science 38: 68-69.

Hutton, E.M. and Gray, S.G. 1967. Hybridization between the
legumes Desmodium intortum, D. uncinatum and D. sandwicense.
Journal of the Australian Institute of Agricultural Science
33: 122-123.

Imrie, B.C. 1972a. Effect of seed size on germination and seedling yield of Desmodium. SABRAO Newsletter 4: 85-89.

Imrie, B.C. 1972b. Desmodium. Division of Tropical Pastures,
CSIRO, Australia, Annual Report 1971-72. pp. 40-41.

Imrie, B.C. 1973. Variation in Desmodium intortum. A preliminary
study. Tropical Grasslands 7: 305-311.

Imrie, B.C. 1975. The use of agar tube culture for early selection for nodulation of Desmodium intortum. Euphytica 24:
625-631.

I'Ons, J.H. 1967. Veld improvement in Swaziland through the introduction of a tropical legume. Proceedings of the Grassland Society of Southern Africa 2: 71-73.

Johansen, C. 1978a. Effect of plant age on element concentrations
in parts of Desmodium intortum cultivar Greenleaf. Communications in Soil Science and Plant Analysis 9: 279-298.

Johansen, C. 1978b. Comparative molybdenum concentrations in some
tropical pasture legumes. Communications in Soil Science and
Plant Analysis 9: 1009-1017.

Johansen, C., Kerridge, P.C. Luck, P.E., Cook, B.G., Lowe, K.F.,
and Ostrowski, H. 1977. The residual effect of molybdenum
fertilizer on growth of tropical pasture legumes in a subtropical environment. Australian Journal of Experimental Agriculture and Animal Husbandry 17: 961-968.

Johansen, C., Kerridge, P.C., Merkley, K.E., Luck, P.E., Cook,
B.G., Lowe, K.F., and Ostrowski, H. 1978. Growth and molyb-
denum response of tropical legume/grass swards at six sites
in south eastern Queensland over a five year period. CSIRO,
Division of Tropical Crops and Pastures, Australia, Tropical
Agronomy Technical Memorandum No. 10.

Jones, R.J. 1973. The effect of frequency and severity of cutting
on yield and persistence of Desmodium intortum cv. Greenleaf
in a subtropical environment. Australian Journal of Experimental Agriculture and Animal Husbandry 13: 171-177.

Jones, R.J. 1974. The relation of animal and pasture production
to stocking rate on legume based and nitrogen fertilized subtropical pastures. Proceedings of the Australian Society for
Animal Production 10: 340-343.

Jones, R.J. and Roe, R. 1976. Seed production, harvesting and
storage. In 'Tropical Pasture Research Principles and Methods' (eds.) N.H. Shaw and W.W. Bryan, Commonwealth Agricultural Bureaux, England. Bulletin 51, pp. 378-392.

Jones, R.M. 1969. Mortality of some tropical grasses and legumes following frosting in the first winter after sowing. Tropical Grasslands 3: 57-63.

Jones, R.M. 1973. Seedling death of Desmodium intortum in the Queensland wallum with special reference to potassium chloride fertilizer. Tropical Grasslands 7: 269-275.

Jones, R.M. 1975. Effect of soil fertility, weed competition, defoliation and legume seeding rate on establishment of tropical pasture species in south-east Queensland. Australian Journal of Experimental Agriculture and Animal Husbandry 15: 54-63.

Jones, R.M. 1980. Persistence of Greenleaf desmodium in established pastures. Tropical Grasslands 14: 123-124.

Jones, R.M. and Evans, T.R. 1977. Soil seed levels of Lotononis bainesii, Desmodium intortum and Trifolium repens in subtropical pastures. Journal of the Australian Institute of Agricultural Science 43: 164-166.

Jones, R.M. and Rees, M.C. 1973. Farmer assessment of pasture establishment reliability in the Gympie district, south-east Queensland. Tropical Grasslands 7: 219-222.

Kemp, D.R. 1976. Observations on the time of sowing and establishment of Lotononis bainesii and Desmodium uncinatum in the Taree district, Australia. Tropical Grasslands 10: 25-32.

Kerridge, P.C., Cook, B.G. and Everett, M.L. 1973. Application of molybdenum tri-oxide in the seed pellet for subtropical pasture legumes. Tropical Grasslands 7: 229-232.

Kerridge, P.C. and Everett, M.L. 1975. Nutrient studies on some soils from Eungella and East Funnel Creek, Mackay hinterland, Queensland. Tropical Grasslands 9: 219-228.

Keya, N.C.O. 1974. Grass/legume pastures in western Kenya. 2. Legume performance at Kitale, Kisii and Kakemaga. East African Agriculture and Forestry Journal 39: 247-257.

Keya, N.C.O. 1976. The role of Desmodium uncinatum (Jacq.) DC. in the improvement of uncultivated grassland of western Kenya. Ph.D. Thesis, University of Nairobi.

Keya, N.C.O. and Eijnatten, C.L.M. 1975a. Studies on oversowing of natural grasslands in Kenya. 1. The effects of seed threshing, scarification and storage on the germination of Desmodium uncinatum (Jacq.) DC. East African Agriculture and Forestry Journal 40: 261-263.

Keya, N.C.O. and Eijnatten, C.L.M. 1975b. Studies on oversowing of natural grasslands in Kenya. 2. Effects of seed inoculation and pelleting on the nodulation and productivity of Desmodium uncinatum (Jacq.) DC. East African Agriculture and Forestry Journal 40: 351-358.

Keya, N.C.O. and Kalangi, D.W. 1973. The seeding and superphosphate rates for the establishment of Desmodium uncinatum by oversowing in uncultivated grasslands of Western Kenya. Tropical Grasslands 7: 319-325.

Keya, N.C.O., Olsen, F.J. and Holliday, R. 1972a. The establishment of desirable grass/legume mixtures in unimproved grassland by planting root splits. East African Agricultural and Forestry Journal 37: 220-223.

Keya, N.C.O., Olsen, F.J. and Holliday. R 1972b. Comparison of seed-beds for oversowing a Chloris gayana Kunth/Desmodium uncinatum (Jacq.) DC. mixture in Hyparrhenia grassland. East African Agriculture and Forestry Journal 37: 286-293.

Kitamura, Y., Nishimura, S. and Tanaka, S. 1976. Studies on mixed cultivation of tropical legume and grass. Part II. Effects of light intensity, inoculation and nitrogen application on the early growth of desmodium (Desmodium intortum). Journal of the Japanese Society of Grassland Science 22: 116-120.

Kretschmer, A.E. 1970. Production of annual and perennial tropical grasses in Florida. Proceedings of the XI International Grasslands Congress, Surfer's Paradise pp. 149-153.

Kretschmer, A.E. 1971. New legumes for the Latin American Tropics. Agricultural Research Center, Fort Pierce, University of Florida. Mimeo. Report 71-73.

Kretschmer, A.E., Brolmann, J.B., Snyder, G.H., and Gascho, G.J. 1973. Production of six tropical legumes each in combination with three tropical grasses in Florida. Agronomy Journal 65: 890-892.

Kretschmer, A.E., Brolmann, J.B., Snyder, G.H., and Coleman, S.W. 1976. 'Florida' carpon desmodium (Desmodium heterocarpon (L.) DC) a perennial tropical forage legume for use in south Florida. Florida Agricultural Experiment Station Circular S-260.

Kretschmer, A.E., Sonoda, R.M., and Snyder, G.J. 1980. Resistance of Desmodium heterocarpon and other tropical legumes to root-knot nematodes. Tropical Grasslands 14: 115-120.

Little, D.A. 1976. Assessment of several pasture species, particularly tropical legumes, for oestrogenic activity. Australian Journal of Agricultural Research 27: 681-686.

Loch, D.S. 1977. Brachiaria decumbens (Signal grass) - A review with particular reference to Australia. Tropical Grasslands 11: 141-157.

Lowe, K.F. 1976. The value of a frost tolerant setaria component in mixed pastures for autumn saved feed in south-eastern Queensland. Tropical Grasslands 10: 89-97.

Lu, S-S. 1969. Introduction of tropical green manuring plants and pastures on 3 red soils of different fertility in W. Dwangtung. Field Crop Abstracts 22: #2240, p. 299.

Mannetje, L. 't and Pritchard, A.J. 1974. The effect of daylength and temperature on introduced legumes and grasses for the tropics and subtropics of coastal Australia. 1. Dry matter production, tillering and leaf area. Australian Journal of Experimental Agriculture and Animal Husbandry 14: 173-181.

Mappledoram, B.D. and Theron, E.P. 1972. Notes on the adaptability of several tropical legumes to different environments in Natal. Proceedings of the Grassland Society of Southern Africa 7: 84-86.

McWhirter, K.S. 1969. Cytoplasmic male sterility in Desmodium. Australian Journal of Agricultural Research 20: 227-241.

Middleton, C.H. 1970. Some effects of grass-legume sowing rates on tropical species establishment and production. Proceedings of the XI International Grasslands Congress, Surfer's

136

Paradise pp. 119-123.

Miller, C.P. and van der List, J.T. 1977. Yield, nitrogen uptake and liveweight gains from irrigated grass legume pasture on a Queensland tropical highland. Australian Journal of Experimental Agriculture and Animal Husbandry 17: 949-960.

Moomaw, J.C. and Takahashi, M. 1962. Forage and beef production from pangolagrass-Desmodium intortum pastures in Hawaii. Agronomy Abstracts p. 95.

Munns, D.N. and Fox, R.L. 1977. Comparative lime requirements of tropical and temperate legumes. Plant and Soil 46: 533-548.

Murphy, W.M., Scholl, J.M., and Baretto, I. 1977. Effects of cutting management on eight subtropical pasture mixtures. Agronomy Journal 69: 662-666.

Naveh, Z. 1971. A synopsis on introduction and selection of grasses and legumes for intensive tropical pastures in Northern Tanzania (1963-65). In "Conference on the intensive management of forage production in the humid tropics." (eds.) J.E. Salette and M. Chenost pp. 14-15.

Naveh, Z. and Anderson, G.D. 1967. Promising pasture plants for Northern Tanzania. IV. Legumes, grasses and grass/legume mixtures. East African Agriculture and Forestry Journal 32: 282-304.

Nel, L.O. and Bosman, S.W. 1979. Beef production potential of Desmodium intortum cv. Greenleaf in the coastal belt to the north of East London. Grassland Society of Southern Africa Proceedings 14:109-110.

Nicholls, D.F., Gibson, T.A., Humphreys, L.R. Hunter, G.D., and Bahnisch, L.M. 1973. Nitrogen and phosphorus response of Desmodium uncinatum on seed production at Mt. Cotton, south eastern Queensland. Tropical Grasslands 7: 243-248.

Noble, A. and Lowe, K.F. 1974. Alcohol soluble carbohydrates in various tropical and temperate pasture species. Tropical Grasslands 8: 179-188.

Norris, D.O. 1972. Seed pelleting to improve nodulation of tropical and subtropical legumes. 4. The effects of various mineral dusts on nodulation of Desmodium uncinatum. Australian Journal of Experimental Agriculture and Animal Husbandry 12: 152-158.

Norris, D.O. 1973. Seed pelleting to improve nodulation of tropical and subtropical legumes. Part 6. The effects of dilute sticker and of bauxite pelleting on nodulation of six legumes. Australian Journal of Experimental Agriculture and Animal Husbandry 13: 700-704.

Ohashi, H. 1973. The Asiatic species of Desmodium and its allied genera (Leguminosae). Ginkgoana - Contributions to the flora of Asia and the Pacific region No. 1. (Academia Scientific Book Inc: Tokyo).

Olsen, F.J. 1973. Effects of cutting management on a Desmodium intortum/Setaria sphacelata mixture. Agronomy Journal 65: 714-716.

Olsen, F.J. and Moe, P.G. 1971. The effect of phosphate and lime on the establishment, productivity, nodulation and persistence of Desmodium intortum, Medicago sativa and Stylosanthes

gracilis. East African Agriculture and Forestry Journal 37:
29-37.

Olsen, F.J. and Tiharuhondi, E.R. 1972. The productivity and botanical composition of some selected grass/legume pasture
mixtures at different seeding rates. East African Agricultural and Forestry Journal 38: 16-22.

Park, R.J. and Minson, D.J. 1972. Flavour differences in meat
from lambs grazed on tropical legumes. Journal of Agricultural Science, Cambridge 79: 473-478.

Park, S.J. and Rotar, P.P. 1968. Genetic studies in Spanish Clover, Desmodium sandwicense E. Mey. 1. Inheritance of flower
color, stem color, and leaflet markings. Crop Science 8:
467-470.

Partridge, I.J. 1975. The improvement of mission grass (Pennise-
setum polystachon) in Fiji by top dressing superphosphate and
oversowing a legume (Macroptilium atropurpureum). Tropical
Grasslands 9: 45-52.

Partridge, I.J. 1980. The effect of grazing and superphosphate on
the spread of a naturalized legume, Desmodium heterophyllum,
on hill land in Fiji. Tropical Grasslands 14: 63-68.

Pritchard, A.J. and Gould, K.F. 1964. Chromosome numbers in some
introduced and indiginous legumes and grasses. CSIRO, Division of Tropical Pastures, Australia. Technical Paper No.
2.

Rand, J.R. and Braithwaite, B.M. 1975. Control of the weevil Amnemus quadrituberculatus in tropical legume pastures with
dieldrin and heptachlor presowing soil treatments. Australian Journal of Experimental Agriculture and Animal Husbandry
15: 545-549.

Rayment, G.E., Compton, B.L., and McDonald, R.C. 1979. Relationships between soils, current nutrient status and persistence
of Greenleaf desmodium at Tinana in south-eastern Queensland.
Tropical Grasslands 13: 53-62.

Rees, M.C., Jones, R.M., and Roe, R. 1976. Evaluation of pasture
grasses and legumes grown in mixtures in south-east Queensland. Tropical Grasslands 10: 65-78.

Riveros, F. and Wilson, G.L. 1970. Responses of a Setaria sphacelata - Desmodium intortum mixture to height and frequency of
cutting. Proceedings of the XI International Grasslands
Congress, Surfer's Paradise pp. 666-668.

Roberts, C.R. 1974. Some problems of establishment and management
of legume based tropical pastures. Tropical Grasslands 8:
61-67.

Roberts, C.R. 1979. Some common causes of failure of tropical le-
gume/grass pastures on commercial farms and suggested reme-
dies. In "Pasture Production in Acid Soils of the tropics".
(eds.) P.A. Sanchez and L.E. Tergas CIAT, Cali, Colombia pp.
123-137.

Roberts, O.T. 1970. A review of pasture species in Fiji. II. Legumes. Tropical Grasslands 4: 213-222.

Rotar, P.P. and Chow, K.H. 1971. Morphological variation and interspecific hybridization among Desmodium intortum, Desmodium
sandwicense, and Desmodium uncinatum. Hawaii Agricultural

Experiment Station, University, Technical Bulletin No. 82.

Rotar, P.P. and Urata, U. 1967. Cytological studies in the genus *Desmodium*; some chromosome counts. American Journal of Botany 54: 1-4.

Rotar, P.P., Park, S.J. Bromdep, A., and Urata, U. 1967. Crossing and flowering behavior in Spanish clover, *Desmodium sandwicense* E. Mey., and other *Desmodium* species. Hawaii Agricultural Experiment Station, University of Hawaii, Technical Progress Report No. 164.

Russell, J.S. 1976. Comparative salt tolerance of some tropical and temperate legumes and tropical grasses. Australian Journal of Experimental Agriculture and Animal Husbandry 16: 103-109.

Schultze-Kraft, R. and Giacometti, D. 1979. Genetic resources of forage legumes for the acid, infertile savannas of tropical America. In "Pasture Production in Acid Soils of the Tropics" (eds.) P.A. Sanchez and L.E. Tergas CIAT, Colombia pp. 55-64.

Schubert, Bernice G. 1963. *Desmodium*: Preliminary studies IV. Journal of the Arnold Arboretum, XLIV: 284-297.

Shaw, K.A. and Quinlan, T.J. 1978. Dry matter production and chemical composition of Kenya white clover, white clover and some tropical legumes grown with *Pennisetum clandestinum* in cut swards on the Evelyn Tableland of North Queensland. Tropical Grasslands 12: 49-57.

Skerman, P.J. 1977. Tropical Forage Legumes. FAO Plant Production and Protection Series No. 2 FAO: Rome.

Smith, A. 1977. The evaluation of tropical pasture species in the Transvaal. Proceedings of the Grassland Society of Southern Africa 12: 29-31.

Snyder, G.H., Kretschmer, A.E. and Sartain, J.B. 1978. Field response of four tropical legumes to lime and superphosphate. Agronomy Journal 70: 269-273.

Stobbs, T.H. 1969. Beef production from pasture leys in Uganda. Journal of the British Grassland Society 24: 81-86.

Stobbs, T.H. 1971. Production and composition of milk from cows grazing Siratro (*Phaseolus atropurpureus*) and Greenleaf desmodium (*Desmodium intortum*). Australian Journal of Experimental Agriculture and Animal Husbandry 11: 268-273.

Stobbs, T.H. and Imrie, B.C. 1976. Variation in yield, canopy structure, chemical composition and in vitro digestibility within and between two *Desmodium* species and interspecific hybrids. Tropical Grasslands 10: 99-106.

Stobbs, T.H. and Thompson, P.A.C. 1975. Milk production from tropical pastures. World Animal Review 13: 27-31.

Sweeney, F.C. and Hopkinson, J.M. 1975. Vegetative growth of nineteen tropical and subtropical pasture grasses and legumes in relation to temperature. Tropical Grasslands 9: 209-217.

Teitzel, J.K., Abbott, R.A., and Mellor, W. 1974. Beef cattle pastures in the wet tropics - 3. Queensland Agricultural Journal 100: 185-189.

Tetteh, A. 1972. Comparative dry matter yield patterns of grass/legume mixtures and their pure stands. Ghana Journal of Agricultural Science 5: 195-199.

Thomas, D. 1975. Growing leguminous pastures on the Lilongwe Plain. World Crops 7: 285-260.

Thomas, D. 1976a. Effects of close grazing or cutting on the productivity of tropical legumes in pure stand in Malawi. Tropical Agriculture 53: 329-333.

Thomas, D. 1976b. Effects of close grazing on the productivity and persistence of tropical legumes with rhodes grass in Malawi. Tropical Agriculture 53: 321-327.

Tudsri, S. and Whiteman, P.C. 1977. Effects of initial and maintenance phosphorus levels on the establishment of four legumes oversown into Setaria anceps swards. Australian Journal of Experimental Agriculture and Animal Husbandry 17: 629-636.

Valdez, R.B. 1975. Some studies on nematodes of forage and pasture crops in the Philippines. Proceedings of the Fifth Scientific meeting of the Crop Science Society of the Philippines, May 1975 pp. 69-74.

Vallis, I. and Jones R.J. 1973. Net mineralization of nitrogen in leaves and leaf litter of Desmodium intortum and Phaseolus atropurpureus mixed with soil. Soil Biology and Biochemistry 5: 391-398.

Wendt, W.B. 1971. Effects of inoculation and fertilizers on Desmodium intortum at Serere, Uganda. East African Agriculture and Forestry Journal 36: 317-321.

White, R.E. 1972. Studies on mineral ion absorption by plants. 1. The absorption and utilization of phosphate by Stylosanthes humilis, Phaseolus atropurpureus and Desmodium intortum. Plant and Soil 36: 427-447.

Whiteman, P.C. 1968. The effects of temperature on the vegetative growth of six tropical legumes species. Australian Journal of Experimental Agriculture and Animal Husbandry 8: 528-532.

Whiteman, P.C. 1969. The effects of close grazing and cutting on the yield, persistence and nitrogen content of four tropical legumes with rhodes grass at Samford, south-eastern Queensland. Australian Journal of Experimental Agriculture and Animal Husbandry 9: 287-294.

Whiteman, P.C. 1970. Seasonal changes in growth and nodulation of perennial tropical pasture legumes in the field. II. Effects of controlled defoliation levels on nodulation of Desmodium intortum and Phaseolus atropurpureus. Australian Journal of Agricultural Research 21: 207-214.

Whiteman, P.C. 1971. Distribution and weight of nodules in tropical legumes in the field. Experimental Agriculture 7: 75-85.

Whiteman, P.C., Bohoquez, M., and Ranacou, E.N. 1974. Shading tolerance in four tropical pasture legume species. In "Biological and physiological aspects of the intensification of grassland utilization." Proceedings of the XII International Grasslands Conference, Moscow (Conference Edition) pp. 402-407.

Whiteman, P.C. and Lulham, A. 1970. Seasonal changes in growth and nodulation of perennial tropical pasture legumes in the field. 1. The influence of planting date and grazing and cutting on Desmodium uncinatum and Phaseolus atropurpureus. Australian Journal of Agricultural Research 21: 195-206.

Whitney, A.S. 1970. Effects of harvesting interval, height of cut and nitrogen fertilization on the performance of Desmodium intortum mixtures in Hawaii. Proceedings of the XI International Grasslands Conference, Surfer's Paradise, Australia pp. 632-636.

Whitney, A.S. and Green, R.E. 1969. Legume contributions to yields and compositions of Desmodium spp. - pangola grass mixtures. Agronomy Journal 61: 741-746.

Whitney, A.S., Kanehiro, Y., and Sherman, G.D. 1967. Nitrogen relationships of three forage legumes in pure stands and in grass mixtures. Agronomy Journal 59: 47-50.

Williams, C. and Yunus, M.M. 1975. A biological study of Desmodium triflorum. Malaysian Agricultural Research 4: 163-172.

Wilson, G.L. 1976. Setaria-desmodium cutting trial. Tropical Grasslands 10: 231.

Younge, O.R. and Plucknett, D.L. 1965. Beef production with heavy phosphorus fertilization in infertile wetlands of Hawaii. Proceedings of the IX International Grasslands Congress, Sao Paulo. 959-963.

Younge, O.R., Plucknett, D.L., and Rotar, P.P. 1964. Culture and yield performance of Desmodium intortum and Desmodium canum in Hawaii. Hawaii Agricultural Experiment Station Bulletin 59.

4
Stylosanthes

R. L. Burt, D. G. Cameron,
D. F. Cameron, L.'t Mannetje
and J. Lenne

INTRODUCTION

Stylosanthes is a relatively small genus of tropical legumes most species of which are native to South and Central America and the Caribbean. Within their native areas many of these plants have long been valued by the local pastoralists (Skerman, 1970; Pittier, 1944). Only recently have attempts been made to domesticate them; for example, S. scabra (referred to as S. diarthra by Pittier) has only just yielded its first commercial cultivar. The domestication of tropical pasture legumes is itself a very recent phenomenon. In the 1930s the need for pasture legumes in the tropics was recognized, but sources were held to be in doubt. William Davies (1933) considered the breeding of indigenous plants and the extension of the use of temperate species to be alternatives to the use of imported tropical species. It was eight years later before Schofield (1941) was able to state that "Tropical legumes are satisfactory under coastal conditions in north Queensland but temperate legumes are markedly unsuccessful." It was not until the early 1960s before definite evidence of the value of tropical legumes in drier areas was forthcoming (see Gillard, Section I-6).

Thus we are dealing with a very new subject and there are still large gaps in our knowledge. Because of the potential importance of Stylosanthes, emphasis has often been placed on the study of its component species. It has already furnished over one third of all of the cultivars used as sown pasture legumes in tropical and sub-tropical regions.

In this overview the knowledge assembled to date will be used to indicate which avenues of research have been most rewarding and what remains to be done. Many of the principles and considerations involved should apply equally well to other, more poorly known genera.

DOMESTICATION OF STYLOSANTHES

Published reports indicate that the first species of Stylo-

141

santhes to be domesticated. was <u>S. guianensis,</u> which was used as a cover crop in the plantations of Malaysia. The first species to attract interest as a pasture legume was Townsville stylo, <u>S. humilis,</u> an annual from South and Central America. After its accidental introduction into north Queensland, Australia around the turn of the century it spread throughout the area and was regarded as "a sample of leguminous weed...of high feeding value." It found local advocates who recommended it to dairy farmers and locally favored lines and mixtures were distributed to graziers and sold commercially. By the mid 1930s its merits were recognized, but it was not until the 1960s that evidence of the value of this species was documented (Norman and Arndt, 1959; Shaw, 1961) and specific cultivars were made commercially available.

Townsville stylo had, however, evoked considerable interest before this time. McTaggart (1937) wrote that "Encouraged by the success of the annual <u>Stylosanthes sundaica</u>[1] in improving or providing pasture...the Division of Plant Industry sought, by correspondence overseas, a perennial species of <u>Stylosanthes</u> which in tropical Australia might prove to be a true substitute for lucerne (<u>Medicago sativa</u>) as a pasture legume. The search resulted in the introduction, on October 6, 1933, of the perennial species <u>Stylosanthes guyanensis</u>[2] ...from Brazil."

Widely tested throughout northern Australia, <u>S. guianensis</u> was found to be particularly well adapted to acid, infertile, phosphorus deficient soils in wet areas. Schofield, working at South Johnstone, proved its value as a pasture species (see Teitzel and Middleton, Section I-7). It was subsequently given the cultivar name of "Schofield" in 1966.

The success of the species <u>S. humilis</u> for dry tropical areas and <u>S. guianensis</u> for more humid regions stimulated a great deal of interest in the genus. It should be noted, however, that this success was based on a very limited range of plant material. Further introduction by correspondence followed. It was not until 1947-48 that a collecting mission was undertaken to specifically provide material and even then the main concern was with the subtropics (Hartley 1949). One of the introductions collected on this mission, subsequently to be made available as <u>S. guianensis</u> var. intermedia cv. Oxley, proved to be quite different from the cultivar Schofield; it was well adapted to drier and colder subtropical areas and was very different morphologically. The wide ecological and agronomic variation within a <u>Stylosanthes</u> species was then recognized.

By the 1960s many accessions were obtained by correspondence and, although the emphasis had been heavily placed on <u>S. guianensis</u> and <u>S. humilis</u>, there was a scattering of introductions from other species. Various collecting trips were made to tropical areas and the existing collections were augmented; several Australian organizations held collections of the total available range of material. To avoid a piecemeal approach to evaluation, these

---

[1]now known to be <u>S. humilis</u>.
[2]now <u>S. guianensis</u>.

collections were assembled by R.J. Williams and transferred to Townsville, Queensland, Australia. Little or nothing was known about some of the species and some accessions were unnamed. Because these species were new and of unknown performance it was not even clear which characters should be measured. Taxonomic descriptions by themselves would not be adequate as the collection contained markedly different forms of the same species.

Beginning in 1967 an experiment was carried out which was designed to meaningfully describe the collection to all concerned with its utilization. For each accession, the usual agronomic characteristics were measured - yield, flowering time, persistence, and so forth. In addition, ther morphological information was collected, which could be used to distinguish lines - shape of leaflets, degree of viscidity, et cetera. The resultant information was analyzed using a computer-based classification program and the 154 accessions were divided into "morphological-agronomic" (M-A) groups (Burt et al., 1971). To a large extent the taxonomy of the species was maintained but the M-A groups proved far more informative than the species (see Section III). Several species appeared to be adapted to dry tropical environments and a considerable amount of useful agronomic variation was found to exist in S. guianensis. Work since 1971 has resulted in the commercial release of "new" species - S. hamata cv. Verano and S. scabra cvs. Seca and Fitzroy - for dry areas and S. guianensis cvs. Cook, Endeavour and Graham for more humid regions. Other species in the collection were found to have agronomic potential and may be made available in the future. A coordinated approach to the development of this collection promoted a quick, efficient utilization of this resource (see also Multidisciplinary Approaches, Section III, 7).

Recent work at other Tropical Research Centers appears certain to result in the release of additional new cultivars. The work at CIAT[3] (Cali, Colombia) has demonstrated that S. capitata is well adapted to the extremely infertile, very acid soils found on the Llanos and Cerrado areas of South America. In Florida and Antigua "new" forms of S. hamata were found which thrive in alkaline soils, situations which at one time appeared to be unsuited to Stylosanthes. The potential of Stylosanthes is far from exhausted.

Past introduction and evaluation programs tended to adopt a rather parochial attitude and have concentrated heavily on species of immediate promise in the area concerned. Such is understandable but it can have an unfortunate outcome as may be seen in the following examples relating to the domestication of S. hamata, a species long known to have agronomic potential. S. hamata was first introduced into Australia in the 1930s and, after very favorable comments from McTaggart (1937), its significance was lost, presumably because it failed to survive in the areas in which it was tested. This is not surprising as the introduction originated

---

[3]CIAT = Centro Internacional de Agricultura Tropical; Cali, Colombia

from a region of strongly alkaline soils of Guadeloupe and was
ill-adapted to the acid soils on which it was originally tested.
Had this accession been tested on alkaline soils, the current interest in S. hamata (virtually based on an almost chance introduction by Atkinson) could well have been aroused in the 1930s; the
cultivar Verano and the promising lines in Florida, Antigua and
Columbia could perhaps have been developed at a much earlier time.
International collaboration and cooperation in the collection and
evaluation of tropical pasture legumes is obviously necessary.

THE RESOURCE POTENTIAL

By tropical standards, Stylosanthes is a small genus: it
contains only about 30 named species (Mohlenbrock, 1957) as opposed to 300 to 500 for Desmodium. Nor does it have the two centers of diversity found in Desmodium; there are only three species
of Stylosanthes native to areas outside South and Central America
and the Caribbean: S. sundaica (sometimes considered to be con-
specific with S. humilis) is from Malaysia, S. fruticosa from Africa and India, and S. erecta from Africa.
The remaining species are found over a very wide geographical
range extending from beyond 30°S to over 40°N--from the Argentine
to New Jersey, U.S.A. Within this area they are found on soils
ranging from pure sands to heavy clays and on soils with pH's
ranging from 4 to over 9. They are found in areas which receive
less than 400 mm of rain annually to those with more than 4000 mm
and in areas with maximum temperatures of more than 40°C to those
with minimum temperatures below zero.
The largest collections (gene or germplasm banks) of Stylosanthes at the present time are held by CIAT (in Colombia) and
CSIRO (in Australia). Although one can use the descriptions of
these collections as indicators of their adequacy there are many
reasons why these must be regarded purely as indicators. Other
institutions are now involved; EMBRAPA (Brazil), for example, is
establishing substantial collections (Giacometti, 1979) and is
publishing well documented information about them (for example, De
Sousa Costa and Ferreira, 1977). Although one can measure the
size of a collection one cannot measure its quality; relatively
small collections may contain potentially very valuable material
which is not held elsewhere. Finally, it should be noted that
researchers are in no position to begin to define 'quality'; a
species of no particular interest at the present time in a given
location may quickly prove to be of tremendous agronomic importance. Details of the CIAT and CSIRO Stylosanthes collections can
be used to indicate many important features about the development
of tropical pasture legume germplasm resources (Table 4.1).
The first point to note is the size of the collections. In
total they number less than 2000 accessions, about 1/10 of the
size of the World Phaseolus collection (Anon., 1978). A second
feature is the species representation; 4 species are not represented at all and a further 14 have less than 10 accessions each.

TABLE 4.1
Numbers of _Stylosanthes_ Accessions Held by CSIRO and CIAT[1]

Species	CSIRO 1969[2]	CSIRO 1978[3]	CIAT 1978[4]	Total 1978
S. angustifolia	2	4	7	11
S. biflora	0	2	0	2
S. bracteata	0	0	5	5
S. calcicola	1	2	4	7
S. capitata	3	23	51	74
S. cayennensis (S. hispida)	1	1	0	1
S. erecta	3	15	1	16
S. figueroae	0	0	0	0
S. fruticosa	34	72	4	76
S. guianensis	121	257	438	695
S. hamata	11	154	63	217
S. hippocampoides	1	1	0	1
S. humilis	49	80	113	193
S. ingrata	0	3	5	8
S. leiocarpa	1	3	1	4
S. macrocarpa	1	1	0	1
S. macrocephala	0	0	1	1
S. macrosoma	0	1	0	1
S. mexicana	0	2	1	3
S. montevidensis	3	7	38	45
S. nervosa	0	0	0	0
S. pilosa	0	0	1	1
S. scabra	25	49	115	164
S. sericeiceps	0	0	0	0
S. subsericea	17	26	3	29
S. sundaica	1	4	2	6
S. sympodialis	0	6	8	14
S. tuberculata	0	0	0	0
S. viscosa	11	29	67	96
S. sp. aff. hamata	0	8	0	8
S. spp.	0	170	49	219
	285	920	978	1898

[1] It is estimated (1978) that about 120 accessions are common to both collections.

[2] from Edye et al., 1974.

[3] from list compiled by R.J. Williams (pers. comm. 1978).

[4] from CIAT (1978).

Many Stylosanthes species contain a great deal of variation. Some groups of plants now taxonomically classified as one species (Mohlenbrock, 1957) are so heterogeneous that they were formerly classified as several species. If one may arbitrarily deem that any collection containing less than ten accessions per species gave an inadequate representation of that species, then fourteen of the twenty-five species (Table 4.1) are poorly covered. Certainly a species could not be said to be of "no agronomic interest" if it was so narrowly represented.

The reverse situation would appear to be the case with S. guianensis which alone comprises more than 40 percent of the accessions in the CIAT and CSIRO collections (Table 4.1). Unfortunately this species is not as well represented as the numbers of accessions in these collections might make it appear. S. guianensis occurs over a wide geographic range which has been arbitrarily split into four regions (Table 4.2). First is South America, excluding Peru and Bolivia west of the Andes. The latter, our second region, is known to be floristically different; with the Galapagos Islands it is the only region in which S. sympodialis occurs. It is one of the centers of diversity for a number of plants, among which are primitive cottons. Central America constitutes a third region. It appears to be the area in which one of the most desirable forms of S. guianensis occurs. Finally, there is Mexico, a country known to be the center of diversity for a number of crop species and the only area in which some varieties of S. guianensis occur ('t Mannetje, 1977).

It is immediately apparent that present collections are very biased away from certain areas. It is most unlikely that there is even a reasonable representation of the variation to be found in S. guianensis, the best collected species. Two varieties are not represented at all ('t Mannetje, 1977).

TABLE 4.2
Regions from which the CSIRO/CIAT Germplasm Collections of S. guianensis Were Assembled (1978)

Region	Numbers of Accessions CSIRO	CIAT
1. South America, excluding Zone 2.	221	393
2. Equador, Peru, Bolivia, west of Andes	1	0
3. Central America	27	45
4. Mexico	8	0
	257	438

TAXONOMY

According to Hutchinson (1964) <u>Stylosanthes</u>, together with Arachis, Zornia, Chapmannia, and Pachecoa, belongs to the tribe Stylosanthea of the family Fabaceae (Papilionaceae). Chapmannia (1 herb from Florida) and Pachecoa (3 shrubs from tropical America) are not used as pasture legumes, but the other genera are. Zornia has a pod of three or more articulations and 2- to 4-foliate leaves, whereas Stylosanthes has one or two articulations and 3-foliate leaves. Arachis has two pairs of leaflets (rarely 3-foliate) and the fruit matures below the ground.

The genus Stylosanthes, established by Swartz (1788), has been revised three times (Vogel, 1838; Taubert, 1891; and Mohlenbrock, 1957). Vogel's division of the genus into sections has been maintained throughout. Species in the section Styposanthes possess a rudimentary secondary floral axis (axis rudiment) and two inner bracteoles, while those in the section Stylosanthes lack an 'axis' rudiment and have only one inner bracteole. However, the axis rudiment is frequently cauducous in some species, e.g. S. hamata, S. sericeiceps and S. sundaica. In addition, Blake (1924) and Mohlenbrock (1957) have stated that S. sericeiceps has only one inner bracteole. Despite these exceptions the recognition of sections is supported by cytological (Cameron, 1967), morphological, and distributional evidence. All known polyploids and all species occurring away from the center of origin, except the doubtful African species S. suborbiculata, belong to the section Styposanthes.

Species of the genus Stylosanthes are morphologically difficult to distinguish from one another. The flowers are uniform and inflorescence differences occur only in the number of inner bracteoles (1 or 2), in the presence or absence of an axis rudiment, in the number of articulations of the pod, in the shape and hairiness (indument) of the upper articulation and in the length and curvature of the beak of the pod (hardened style). Vegetative differences are generally slight except for the shape of leaflets, the width of the floral bract and the presence or absence of bristles on the stem in some species. While the morphological similarities between species are not indicative of similar agronomic behavior, Burt et al. (1971) found that a grouping of Stylosanthes accessions on the basis of morphological plus agronomic attributes was valuable in predicting their usefulness as pasture legumes.

Most tropical pasture legumes have a twining habit, which limits their adaptation to grazing ('t Mannetje et al., 1979). Stylosanthes species on the other hand are all sub-erect to erect herbs or small shrubs and either annual or perennial with copious seed production. Several Stylosanthes species are well adapted to grazing either by having a prostrate habit (e.g. S. humilis), or by having woody stems which are not grazed (e.g. S. scabra). The non-dehiscent pod (not splitting open at maturity) is an advantage for seed harvesting and regulates germination of annual species to some extent by preventing germination as a result of light showers at the beginning of the wet season (Holm, 1973).

Vogel (1838), Taubert (1891), and several other authors (e.g.

Hassler, 1919) recognized varieties in several species. However, Mohlenbrock (1957) abolished all varieties and combined some species. Where true variation existed, taxonomic "lumping" meant a loss of taxonomic or agronomic information or both to economic users. Several varieties of S. guianensis have since been identified ('t Mannetje 1977). Recent examination of herbarium material indicated the need for additional taxonomic research to subdivide other species and to reduce some species to varieties within species. Seed protein pattern interpretations (Robinson and Megarrity, 1975) and the new species and varieties proposed by De Sousa Costa and Ferreira (1977) support this viewpoint. Whereas some of the proposed classifications can be accommodated within species or varieties described earlier, the confusion in the communication of information indicates that the taxonomy of Stylosanthes is in urgent need of revision.

Such a revision can only be done satisfactorily after extensive genetical and cytological studies have been completed. This should be done on more species than those presently available for in vivo research. A first requirement is to systematically collect seed and herbarium specimens from ecological niches and species poorly represented in existing germplasm banks and herbaria.

ADAPTATION

Climatic Adaption

This discussion contains a brief consideration of the climatic adaptation of the various species and their forms as represented in our collections. Current collections, however, are far from complete and the adaptational range of some species is underestimated. If one is to formulate introduction programs for improving tropical pastures it is necessary to have some indication of the true adaptational range of the species of interest (Burt and Reid, 1976). Such information can be inferred from the literature by noting the frequency with which the species occurred in different types of climate. Climatic adaptation of some of the presently more agronomically interesting species are given in Table 4.3.

Table 4.3 indicates that the species differ markedly in their range of climatic adaptation. S. viscosa has a wide climatic range coupled with a wide geographical range (from Paraguay to northern Mexico). S. hamata has a wide climatic range but is limited geographically; it is rarely found outside the Caribbean and the adjoining mainland coasts. S. guianensis is relatively limited climatically--it is rarely found outside the wetter areas--but has a wide geographical range. S. sympodialis is limited both climatically and geographically.

Morphological Adaptation

From the scattered observations which are available it would appear that various morphological features of Stylosanthes sp. tend to favor their success as pasture species. The very factors

TABLE 4.3
Distribution of Various _Stylosanthes_ spp. in Relation to Climate (Based on Data Presented by Burt and Reid, 1976)

| | Climate (After Reid, 1973) | | | | | | Sub- | | |
	Wet tropical highlands	Equatorial extended rainfall	Tropical extended rainfall	Tropical seasonal rainfall	Tropical semi arid	Hot, sub-tropical desert	tropical seasonal rainfall	Dry tropical highlands	Frosty tropical highlands
S. capitata	X[1]			X	XX			XX	
S. mexicana		XXX		X				X	XXX
S. erecta		XXX		XX					
S. scabra	XXX	XXX	XXX	XX			X	XXX	X
S. sympodialis					XXX	XXX			
S. fruticosa				X	XXX		XX	XXX	XX
S. subsericea	XXX			X					X
S. hamata	X	X	XXX	XXX	XXX	XXX	XXX	XX	
S. guianensis	XX	XXX	XXX	XXX	XXX	XXX	XX	X	
S. viscosa	XXX	XXX	XXX	XXX	XXX	XXX	XX	XX	X
S. humilis	XXX	XX	XXX	XXX	X	X	XX	X	X

[1] "X" signifies present, "XX" common, "XXX" frequent.

which tend to separate Stylosanthes from most of the other legume genera which have yielded tropical pasture plants may have agronomic significance. Unlike most of the other genera, Stylosanthes consists entirely of herbs or small shrubs. Usually they have a 'crown' of growing points near the soil's surface which provides the plant with resistance to the grazing animal, and to fire and frost, in contrast to climbing plants which are often susceptible to all three factors. Unlike many of the other genera, Stylosanthes has indehiscent seed, i.e. seeds are retained in the pod. The pod helps to regulate dormancy, thus preventing premature germination during false starts of the rainy season. It aids seed dispersal by preventing total digestion by the grazing animal and by being hooked, which enables it to be dispersed on animals' coats. The seed coat itself may be impermeable and this further helps to regulate dormancy (Gardener, 1975). Such seed characteristics, which tend to be most valuable in dry tropical situations; are not found in S. guianensis, a species adapted to wetter areas with longer growing seasons.

Another feature which may favor Stylosanthes in grazed pasture situations is the relative unpalatability of some species at some times of year; Gillard (Section I-6) and Teitzel and Middleton (Section I-7) mentioned how this characteristic favors survival in S. viscosa, S. humilis and S. guianensis. The reason for this is uncertain but has been part attributed (as with other species) to the presence of viscid hairs.

Although the evidence is meager, it appears that Stylosanthes produces a deep rooting system. Talineau (1970) suggested that this places S. guianensis at an advantage over its less-deeply rooting companion grasses and Gardener (1978) noted that dryland perennials developed deeper, less-branched root systems than annuals with the same root weights. Deep root systems may explain the ability of some of the dryland perennials to grow well into the dry season; they extract water and nutrients unavailable to other shallower rooted species. This possibility is presently being investigated (J. Williams, pers. comm. 1979).

Physiological Adaptation

For this discussion the tropical climates of Queensland, Australia are arbitrarily divided into three types: humid tropical, dry tropical and sub-tropical. The main features of such environments are shown in Figure 4.1. The first (Innisfail), humid tropical is wetter and has a longer growing season than the second; variability of rainfall both within and between seasons is often a characteristic of the second (Townsville). In the sub-tropics (Gayndah) daylength is more variable, minimum temperatures can be much lower (sometimes low enough to frost plants) and rainfall may be more equally distributed. Often, as in our example (Gayndah) the combination of lower temperatures and spring rainfall (in this case before December) allow plant growth at this time of year. Plants are differentially adapted to these conditions.

In humid tropical environments three cultivars of S. guianensis proved to be well adapted physiologically. They are resistant

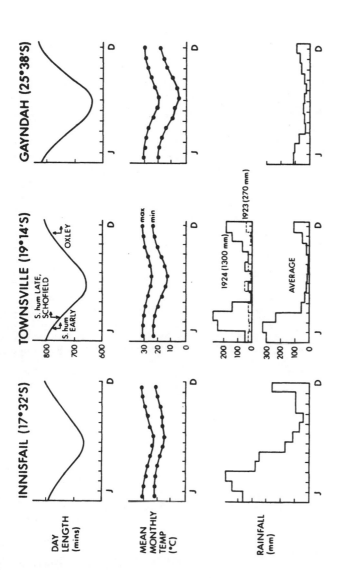

Figure 4.1. Climatic features of Innisfail (humid tropical), Townsville (dry tropical) and Gayndah (sub-tropical), Queensland, Australia.

to flooding but can survive short periods of drought. Like all of the tropical species examined, their germination and growth is favored by high temperatures (Sweeney and Hopkinson, 1975; McIvor, 1976a). Seed production is, however, essential to ensure long term survival, and we find that flower initiation is favored by lower temperatures such as those found toward the end of the growing season (Cameron and 't Mannetje, 1977). Similarly, the end of the growing season is presaged by falling daylengths and these cultivars are short day plants, flowering when daylengths fall below critical levels (see Figure 4.1).

In our dry tropical climate, temperatures and shortening days have a similar effect on the flowering of some forms of S. hamata and the species S. humilis, S. scabra, S. subsericea, and S. viscosa (Cameron and 't Mannetje, 1977). Critical daylengths vary (Figure 4.1); early flowering Townsville stylo initiates flowers when the daylength drops below 13 hours, later flowering varieties have shorter critical daylengths. This is an adaptive characteristic but it can have an unfortunate side effect for pasture systems. Townsville stylo is a determinate plant and vegetative growth virtually ceases after flower initiation. As flowering is fixed by daylength we thus have the situation that, with late starts to the growing season, plants flower and cease growth when they are still very small; they cannot continue to grow even when good follow up rains occur. S. hamata cv. Verano, which is less daylength sensitive and does not have this determinate habit, is not so constrained.

In the sub-tropics, where spring growth is possible, another taxonomic variety of S. guianensis has proved successful. This variety is represented by the cultivar Oxley. Early flowering is again advantageous but in this instance early growth occurs under conditions of increasing daylength. Oxley is a long day plant, flower initiation being promoted by daylengths which exceed 13 hours. It is also different from the other cultivars of S. guianensis in that it is very resistant to drought, fire and overgrazing, all characteristics which suit it to a dry sub-tropical environment.

More detailed physiological studies for the "older" cultivars of S. humilis and S. guianensis may be obtained elsewhere (e.g. Gillard and Fisher, 1978; Rotar et al., 1981). Unfortunately, very little else is available for many of the other species nor can such work sensibly cover them until they are more completely collected. It should be noted that the large physiological differences found among the cultivars of S. guianensis could well exist in other species; in S. hamata there are some forms which are daylength sensitive and others which are not (Cameron and 't Mannetje, 1977) and, even within the naturalized forms of S. humilis found in Australia, there are different temperature requirements for flowering (Schoonhover and Humphreys, 1974).

Much remains to be discovered about the range and limitations of physiological adaptation of Stylosanthes forms and the implications of such findings for tropical pasture improvement.

Edaphic Adaptation

Stylosanthes is usually considered to be a genus which is adapted to infertile soils. Certainly some of the currently available cultivars appear to be more relevant to such conditions than equivalent species from other genera (see, for instance, Teitzel and Middleton, Section I-7 and Robinson, Section III). It would be incorrect, however, to assume that all Stylosanthes species or species forms are usually tolerant of poor soil fertility; some species such as S. sympodialis appear to have a relatively high requirement for soil phosphorus (Section III-1). Impressions of the fertility requirements of this genus are very dependent upon the characteristics of the existing collections, the conditions under which the accessions were tested, and the environments in which the plants were observed to prosper.

Other generalizations must similarly be treated with caution. The suggestion that Stylosanthes is primarily adapted to acid soils stems from work where the genus has proved successful in such situations. In Puerto Rico, where alkaline soils are common, Stylosanthes is considered suited to alkaline situations (Dubey et al., 1974). The opinion may well be an artifact of the existing scientific information network, for by far the greatest number of Stylosanthes citations in the English literature are for plants observed growing in acid soil conditions. It is now known that both acid and alkaline adapted forms exist within the same species (Date et al., 1979). Similarly, the belief that Stylosanthes is unsuited to heavy textured soils may be very much a function of the accessions tested to date; S. sympodialis and S. hamata are both known to thrive on heavy soils and some forms of S. scabra and S. guianensis are currently showing promise in similar situations. Further information on edaphic adaptation is presented elsewhere (Section III).

Agronomic - Ecological Adaptation

Although a plant may be edaphically and climatically adapted there are various reasons why it may be agronomically unacceptable, we will mention three. First, plants shown to be adapted in preliminary evaluation experiments, often in spaced plant rows or in lightly grazed swards, may subsequently succumb to competition from associated species, mismanagement practices (too lenient or too heavy grazing) or to climatically 'bad' years. Fire and flood are also additional stress factors. It is for these reasons that the area thought suitable for Schofield stylo has steadily contracted until its present boundaries were reached in commercial practice. Similarly the area thought appropriate for S. humilis has contracted markedly, particularly when better adapted species were found (Gillard and Fisher, 1978).

Secondly, plants must be suited for the long term requirements of the animal and the agricultural system in which they are to be used; these needs are often difficult to determine in advance. What, for instance, are the advantages of high yielding palatable annuals and short-lived perennials over less palatable perennials that stay green throughout the dry season. In an intensive production system high yield may be a major objective with pastures being replanted at regular intervals. But in extensive systems, the plant's ability to naturalize and persist over long periods of time may be more pertinent than yield.

Thirdly, what are the long term effects of the various plant types on the grazed ecosystem? In tropical pastures based on Townsville stylo, for instance, heavy grazing and the addition of superphosphate are necessary to maintain legume stands. With perennials, high grazing pressures may eliminate the legume which, is better able to withstand grass competition in less heavily grazed, extensive systems.

Conclusions

Compared with their temperate counterparts, our knowledge of even our best known Stylosanthes species is extremely limited. We know little or nothing about other species, such as S. sympodialis, or species forms likely to be of value. This situation is not improving; in 1977 (the last volume of Herbage Abstracts for which an annual index was available), 63 publications contained some information on Stylosanthes. Of these, 52 referred to either S. humilis or S. guianensis and only 11 to the remaining 23 species.

EARLY DOMESTICATED SPECIES

S. humilis[4]

This species is commonly called "Townsville stylo" in Australia and other countries where it is sown. Early publications sometimes referred to this species as S. sundaica, with the common name of "Townsville lucerne." Three cultivars: "Gordon," "Lawson," and "Paterson," which flower at different daylengths, have been identified and released. Black seed color was incorporated into one of them as a marker characteristic. These cultivars were selected from naturally occurring Australian ecotypes growing on acid sandy soils.

Distribution in Australia. Since its introduction into northern Queensland, S. humilis has spread throughout much of the tropics. It is not, however, present in significant amounts in the uplands or at latitudes more than 5°S of the Tropic of Capricorn (Gillard and Fisher, 1978). Its growth is restricted when the mean temperature falls below 23°C (Sweeney and Hopkinson, 1975). It can, however, survive in areas which experience frost in the dry season. S. humilis is not found in wetter areas (>1200 mm; <4 dry months) or in dry regions with less than 600 mm annual rainfall. Almost certainly its exclusion from wetter areas is due to its inability to compete with tall grasses or insufficient time

---

[4]See reviews by Humphreys, 1967 and Gillard and Fisher, 1978.

for its hard seed to break down or both. S. humilis is unsuited to heavy cracking clays or to heavy textured fertile soils. The reasons for this failure are unknown but would be at least part due to its inability to persist in high phosphate soils and excessive competition from native grasses.

Environmental factors affecting use. In broad terms it can be stated (Gillard and Fisher, 1978) that Townsville stylo is adapted to tropical areas with annual rainfall of 800-1200 mm and with dry seasons of more than 4 months. As stated earlier variability of rainfall is important; pastures based on a single genotype of this species may suffer from deterioration or loss of the legume in atypical years. Moisture, high relative humidities, heavy dews, or out of season rains are also important as they greatly reduce the quality of standing hay in the dry season. High humidities in the growing season are also associated with outbreaks of anthracnose disease. An indirect effect of climate is exerted through the companion grass (Gillard and Fisher, 1978). In more tropical dryland areas heavy grazing weakens perennial native grasses which die out and are replaced by annuals. Where the dry season is less severe, for instance in more subtropical areas where temperatures are lower and winter rainfall more frequent, the perennial grasses tend to persist. If left ungrazed, competition from these grasses can eliminate the legume from the sward.

Annual production from Townsville-stylo based pastures. The use of sown pasture species in tropical areas is a relatively new phenomenon; Townsville stylo cultivars were not commercially available until the mid 1960s. Not surprisingly, the documentation of their beneficial effects is relatively restricted. Attempts to use Townsville stylo in conjunction with crops, undersowing and grazing the crop residues, and associated Townsville stylo, have only recently been made. The value of Townsville stylo hay, as a standing crop, has been documented. Norman (1966) showed that cattle gained 89 kg/head during the dry season when grazing a pasture containing 62 percent Townsville stylo and 37 percent annual grasses. When the proportion of grass increased to 77 percent (annuals and perennials) and the proportion of legume fell to 23 percent the equivalent liveweight gain fell to 9 kg/head. In India, where Townsville stylo has been incorporated into native pastures and the dry season hay is cut and fed to dairy cattle, milk production is said to be increased.

In an extensive, continuously grazed situation liveweight gain/ha is probably a more meaningful measure of the value of the legume. Unfortunately, although it seems that Townsville stylo has performed satisfactorily in such varied circumstances as Florida (Kretschmer, 1968) and the dry zones of Nigeria (de Leeuw, 1974), long term grazing experiments rarely appear to have been reported. Liveweight gains at Rodds Bay, Queensland, Australia averaged over a 7-year period, increased as pasture was improved from native pasture (22 kg/ha) to native pasture + Townsville stylo (83 kg/ha) and to native pasture + Townsville stylo + superphosphate (132 kg/ha) (Shaw and 't Mannetje, 1970). At Landsdown,

Queensland, Australia superphosphate added to Townsville stylo pastures increased liveweight production from 115 to 135 kg/ha and increased the conception rate of the cows.

Variation. S. humilis is adventive, not native to Australia, and only limited variation can be found in the Australian populations; such variation is mainly in flowering time, growth habit, seed color (Edye and Cameron, 1975) and in temperature response (Schoonhover and Humphreys, 1974). Most of the other introductions studied have been from Brazil. Like the Australian material these had large leaflets and some (together with other accessions from Colombia) closely resembled those found in Australia (Edye et al., 1975). There is, however, a much greater range of variation within the Brazilian material.

Accessions from Central America are small leaved and are quite variable. Three types have been designated, one of which has flower heads in the stem axils as well as terminally and thick markedly coiled beaks which are very reminiscent of those found on S. subsericea (another annual from Central America) (Edye et al., 1975).

Since these groups were defined, Australian collections of these annual and short-lived perennial species (S. hamata, S. humilis, and S. subsericea) have been expanded and it has become increasingly difficult to separate them. This situation exemplifies a recurrent problem that occurs when a heretofore obscure genus of plants attracts interest because of its newly recognized commercial potential. For example; A review of the literature showed that S. humilis occurred widely in Guatemala (Standley and Steyermark, 1946) over a broad range of climates. Subsequently, a collecting mission found that the species occurred only on strongly alkaline soils (Burt, unpub. data) and that it closely resembled the small leaved form of S. humilis described above. The Guatemalan S. humilis is clearly different from the acid loving forms found in Australia and any attempts made to transfer management technology between these countries must take this difference into account. Taxonomic identification of this material proved difficult as authorities differed on their specific names.

Present status of plant collections. S. humilis may be found in Mexico, Central America, Colombia, Venezuela, and Brazil. Mohlenbrock (1957) initially suggested that the species was adventive to Malaysia where it was known as S. sundaica; later he changed this view. The S. sundaica described by Edye et al. (1975), however, appeared to be quite different from most forms of S. humilis.

Collections to date have been heavily biased towards the larger leaved forms from Brazil and northern South America. Further collection would be useful in searching for anthracnose disease resistance in this species but, in view of the advantages of S. hamata over S. humilis, this should not be given high priority. Central American forms warrant further collection, particularly as they may have some potential for areas with alkaline soils for which there are presently few suitable cultivars.

S. guianensis[5]

This species was commonly referred to as S. guyanensis or S. gracilis. It has many common names: Schofield stylo and stylo (Australia), alfalfa do nordeste and trifolio (Brazil), alfalfa de Brazil (Colombia), and Brazilian lucerne and tarbardillo (Venezuela). Five cultivars are presently available; Cook, Endeavour, Graham, and Schofield are adapted to wet tropical conditions and Oxley is adapted to drier subtropical regions. A sixth cultivar for wetter subtropical areas should be available soon.

Variation. In a recent revision of this species, 't Mannetje (1977) was able to distinguish 6 varieties of S. guianensis. Some varieties at least are variable and contain forms which come from, and are adapted to, quite different climatic and geographic conditions (Burt, Reid, and Williams, 1976). To ease communication for those working with such species, the various forms have been defined in terms of their morphological (M) and agronomic (A) characteristics (Burt et al., 1971, Edye et al., 1974) and will subsequently be referred to as MA groups.

i. S. guianensis var. guianensis. This important variety contains the cultivars Schofield, Cook, Endeavour, and Graham. It is very variable and spans 5 MA groups; Schofield, the oldest cultivar, is from group 7 (using the notation in Edye et al., 1974), Cook and Endeavour are from group 8 and Graham is from group 10.
    The groups differ in a variety of morphological and agronomic features; e.g. groups 5, 7, and 8, are more viscid and tend to have higher dry matter and seed yields than 10 and 14. The groups also tended to come from different climates (Burt, Reid, and Williams, 1976) but most were geographically restricted to South America. Groups 8 and 10 were not, however, so restricted and their prime centre of distribution would appear to be Central America. This area has been poorly collected and in view of the potential of groups 8 and 10 in provision of the new cultivars, further collection in this region is obviously warranted.

ii. S. guianenesis var. gracilis. This variety consists of erect, sparsely branched plants, members of which have exhibited poor persistence and poor vigor in field experiments. Two M/A groups have been nominated. The first (group 11) has long, linear leaflets and smooth, blue-green stems; the specimens examined were from Venezuela and French Guiana. The second group, 13, has sword-shaped (ensiform) leaflets of a pale green color; specimens examined were from Sao Paulo in southern Brazil.

III. S. guianensis var. intermedia. This variety contains only one M/A group (6). Plant habit varies from semi-erect to prostrate and the 'crown' of the plant is located very closely to the soil surface. Plants are highly persistent but of only medium

---

[5]Figures 4.2 and 4.3 (see review by Tuley, 1968; Burt and Miller, 1975; Stonard and Bisset, 1970).

158

Figure 4.2. _Stylosanthes guianensis_, flowers, pods, and seeds.

Figure 4.3. _Stylosanthes guianensis_.

vigor; flowering occurs under increasing daylengths usually greater than 12 hours. The cultivar Oxley belongs to this group. Plants examined in the field were all from low temperature regions in southern Brazil, Uruguay, and Paraguay with corresponding low temperature regimes. Annual rainfall in these regions, although relatively low, was well distributed with part of the wet season occurring in the cooler months.

iv. S. guianensis var. robusta. This variety contains two M/A groups, 4 and 16. Like var. intermedia both flower in daylengths greater than 12 hours and both were either subtropical (Sao Paulo, in southern Brazil) or from tropical highland areas (in Bolivia). The groups differ markedly in many characteristics; accessions in group 4 were prostrate and highly persistent whereas those of group 6 were poorly persistent but quite vigorous.

This brief resume has been used to illustrate the wide range of morphological and agronomic variation found in S. guianensis. There are strong indications that the various morphological/agronomic (M/A) groups tend to come from different geographical situations and different climates. The extent to which these groups are adapted to different climates can be partially judged from the next section which deals with the performance of the groups outside of their regions of origin.

Environmental adaptation. Of the six varieties of S. guianensis two were not represented in the collections studied and two others, the varieties gracilis and robusta, have received scant attention. Although gracilis appeared to have little agronomic value, being poorly persistent and low yielding, this was not the case for the more sub-tropical robusta (Edye et al., 1974).

The cultivars Schofield, Cook, Endeavour, and Graham and Oxley have been developed from the varieties guianensis and intermedia, respectively. These cultivars are agronomically useful in climates similar to those from which they were collected. S. guianensis var. guianensis has been subdivided into 5 M-A groups and, of these, group 8 was the most promising (Edye et al., 1976). Although this group was generally adapted to more humid tropical areas further selection within it could be rewarding; differences in flood tolerance (McIvor, 1976b) and flowering time (Edye et al., 1976) could be of value in fitting plants more closely to local conditions. It is relevant to note that Cook and Endeavour, both from group 8, differed in several important respects: Cook was resistant to the anthracnose strains found in Australia (but not in Colombia) and could grow well into the cooler months; Endeavour provided very vigorous early summer (but not winter) growth but was susceptible to anthracnose, overgrazing and flooding.

The variety intermedia did not exhibit such wide variation; attempts are currently being made to produce variation by cross breeding it with other varieties.

Agricultural usage. The uses of Schofield stylo in plantations, crop rotations, and various pasture management systems have been adequately reviewed by Tuley (1968) and Skerman (1977). Teitzel and Middleton (Section I-7) have presented further information in this volume. The cultivar Oxley has had less general use, probably because of its rather specific Rhizobium inoculant requirement, but its value is well documented (Stonard and Bisset, 1970).

Schofield stylo has been well used in humid tropical areas throughout the world; Skerman (1977) quoted green yields (t/ha/annum) of 20 (Brazil), 35 (Zaire), 70 (Madagascar), 42 (Fiji) and 46 (Zambia). When combined with average protein contents of 12-18 percent (dry weight) it was clear that this represented substantial nitrogen fixation. Skerman (1977) quoted an increase in soil nitrogen of from 34 to 55 ppm following the ploughing under of an 18 month old crop. Schofield stylo can be grazed directly or can be made into hay or silage. It has lead to substantial increases in animal production.

Undoubtedly Schofield stylo has been a plant of major significance in tropical areas. The susceptibility of this cultivar to anthracnose poses a major problem, and the search for a resistant form of S. guianensis, or for another species suited to such situations, is of considerable importance.

NEW SPECIES

These species are 'new' only in the sense that concerted attempts to document their value and to domesticate them have only recently been made. Local pastoralists have long appreciated the value of such species as S. hamata (see, for instance, Stehle, 1956) and S. scabra (Tamayo, quoted by Pittier, 1944, referring to S. scabra by the synonym S. diarthra).

S. hamata

This species is found mainly in the Caribbean islands and the adjoining coastlines of South and Central America and Florida, U.S.A. It may be occasionally found in Brazil but does not appear to be native to that country. It has several common names which include "pencil flower" and "Mother Segal" in the West Indies and "tebeneque" in Venezuela. One cultivar, "Verano," which is adapted to neutral to acid soils and to dry to semi-arid tropical conditions, is commercially available.

Variation. In an earlier work (Burt et al., 1971) two M-A groups were needed to encompass the eight accessions studied. These groups were more similar to other species than they were to each other. It was necessary to subdivide each of these groups and to add a third (Edye et al., 1974) when the size of the collection was increased to 11. Since then interest in S. hamata has grown and Institutes such as CIAT (Colombia), the University of Florida (U.S.A.) and the Unversity of the West Indies (in conjunction with IDRC) have contributed greatly to the collections of this species. It is now clear that the groups of accessions are not discrete but rather that these groups merge and overlap to produce a pattern of continuous variation.

S. hamata tends to "merge" with other species and has strong affinities to S. scabra, S. humilis, S. calcicola, and S. subsericea. The genetic relationships of this complex have yet to be determined but almost certainly interspecific hybridization has occurred (Date, Burt, and Williams, 1979). As yet much of this work is unpublished, and it is convenient to use the M-A group system to describe the type of variation encountered and the adaptation of the material described.

S. hamata was arbitrarily divided into three M/A groups, the first two of which (M/A groups 21 and 28) were herbaceous, nondeterminate plants; Group 21 was multi-branched and had small, elliptic leaves, whereas 28 was less densely branched and had much larger, ensiform leaves. Group 24 consisted of a single accession, an erect woody shrub which resembled S. scabra. Recently, additional Group 24 material was introduced from Brazil; all are persistent perennials quite different from the herbaceous forms of S. hamata. They are tentatively listed as S. sp. aff. hamata in Table 4.1.

These groupings were geographically real: Group 21 members were from the Caribbean Islands and Florida, group 28 was from coastal South America, and Group 24 was from inland areas in Central and South America.

Adaptation

Group 21 accessions were widely tested throughout the Australian tropics and subtropics (Edye et al., 1974). They performed very poorly, as there were no effective Rhizobium strains for them at that time ('t Mannetje, 1969). In Florida (dry tropical), and Antigua (subtropical), Group 21 accessions thrived on alkaline soils varying from coarse coral beach sands to relatively heavy clays over a considerable climatic range. Currently some of the more successful accessions are being considered for commercial release.

The cultivar Verano, from Group 28, has proved well adapted to a range of dry tropical conditions in Australia and is capable of good yields in areas considered to be too dry (semi-arid) for Townsville Stylo (Burt et al., 1974, Edye et al., 1974). Unlike Townsville stylo it has shown some promise on heavier textured soils but this awaits confirmation. Verano is poorly adapted to alkaline soils in Florida and Antigua.

Group 24 members showed high persistence and reasonable vigor in early testing stages but have shown little promise in field testing.

Most of the information on the physiology of this species was limited to work with the cultivar Verano. Group 21 accessions were quite variable with at least one of their members being daylength insensitive. Brolmann (1978) noted that this form was moderately tolerant of flooding and possibly also of frost. Verano appeared to be cold sensitive. Edaphic requirements similarly seemed to differ, the Group 21 accessions showed phosphorus response patterns typical of plants from calcareous soils (Jones, 1974) whereas accessions from other M/A groups demanded much less phosphorus for adequate growth.

Agronomic usage. The only commercially available variety of S. hamata, Verano, was formally released in 1973. Insufficient time has elapsed to allow us to set climatic limits for its use. It is gaining ready acceptance in dry tropical situations where it seems to be replacing S. humilis as the most favored tropical legume. Certainly Verano's anthracnose resistance, higher yielding capacity and ability to compete more successfully with grasses is advantageous. The inclusion of Verano in grazed swards has led to substantial increases in animal production. It is gaining wide acceptance and more seed is produced commercially of it than for any other tropical legume cultivar.

Present status of plant collections. At the present time S. hamata is probably our best collected species. There is good representation from mainland areas in Latin America and North America where the species is known to occur and also from several of the Caribbean Islands. Perhaps the biggest deficiency is the lack of collections from the wetter areas in the Caribbean, particularly from acid soils. Such collections would be valuable where the traditionally used Stylosanthes species, S. guianensis, is threatened by anthracnose.

S. scabra

S. scabra is native to northern South America (Bolivia, Brazil, Colombia, Ecuador, and Venezuela). It has several common names such as "Capitan Juan" and "Pata de terecay" (Venezuela). In relevant literature, such as Pittier (1944), S. scabra and S. diarthra are synonomous (Mohlenbrock 1957). "Fitzroy" is a cultivar adapted to drier and cooler subtropical areas than the presently available cultivar, "Seca."

Variation. The early attempts to characterize the variation patterns in this species were based on a collection of 25 accessions of which the majority were from southern Brazil. As with S. hamata, this paucity of numbers resulted in apparently well defined groups of accessions appearing to have geographic reality. Subsequent collections have resulted in the merging of most of these groups (Burt and Williams, 1979). It is still, however, convenient to use these groups for our present purpose.

Four morphological-agronomic groups were recognized which appeared to have developed under different geographic and climatic conditions: Group 17 members were from tropical, semi-arid areas, Group 18 from dry tropical situations, Group 19 from wetter subtropical areas, and Group 30 from tropical areas with somewhat extended growing seasons.

These groups have already been described elsewhere (Burt et al., 1971, Edye et al., 1974) and it should be noted that they differ in both agronomic and morphological characteristics. Perhaps the main feature to be noted is that unlike S. humilis and S. hamata they are all strong woody perennials.

Subsequent introduction programs provided accessions closely linking groups 17, 19, and 30 in a single continuum (Burt and Williams, 1979). The area from which Group 18 accessions were obtained, Pernambuco-Brazil (8°S)-was not resampled, and this group has retained its identity.

Adaptation. From the limited number of accessions tested to date (Burt et al., 1974, Edye et al., 1976) it is apparent that S. scabra is well adapted to a range of tropical and subtropical conditions. In tropical areas with a dry season of only 4 months, a member of Group 18 outyielded S. guianensis and continued to produce some growth through the dry season. In dry tropical and semi-arid tropical areas an accession from Group 17 has persisted well and has had higher yields than those of Verano; it has performed well in areas that experience frost. A third form, representing Group 19, has shown promise in dry subtropical areas. The various forms thus tend to be adapted to climatic areas similar to those where they were collected. It seems plausible to believe that S. scabra accessions from wet tropical areas could be valuable for areas with similar characteristics elsewhere; this has important implications given the anthracnose resistance commonly found in this species.

Edaphically S. scabra seems to be adapted to sandy surfaced, infertile acid soils. It has usually failed on alkaline soils in Antigua and Florida. One accession has shown some persistence on heavy clay soils.

Agronomic usage. Commercial quantities of seed of the cultivar Seca are just becoming available, and results from grazing experiments are limited. These show that Seca is useful in areas somewhat too dry for S. guianensis; at Heathlands, in northern Cape York, Australia (12°S), cattle gained weight for all but the last month of the five month dry season on a Seca-Brachiaria decumbens pasture. On the equivalent S. guianensis cv. Cook-Brachiaria decumbens pasture they gained weight only for the first two months of the dry season (Winter, pers. comm.).

Seca is adapted to tropical areas with variable summer growing seasons and annual precipitation of 700 to 1700 mm. It competes well with a variety of grasses in both heavily and leniently grazed systems. The latter characteristic is absent in S. humilis and possibly in S. hamata. It is resistant to anthracnose. An accession from Group 17, although susceptible to anthracnose, appears to be better adapted than Seca to drier areas and to those with cooler winters and is now available as the cultivar Fitzroy. More recent introductions have shown a remarkable ability to grow in dry, extremely phosphorus deficient areas.

Present status of plant collections. Stylosanthes scabra is clearly a species of major significance to tropical agriculture. It is pleasing to note that collections of this species have improved markedly in recent years; indeed, collections have been made from some Brazilian states in which the occurrence of S. scabra is not listed by Mohlenbrock (1957). We do not, however, have accessions from some of the other states nor have we expanded our original collections (of two) from Pernambuco (8°S), the area which yielded the anthracnose resistant cultivar Seca. There are no collections from Ecuador and Colombia and few from Bolivia.

S. capitata

At the present time there are no commercially available varieties of this species nor can we find references to its local usage in areas to which it is native. The species thrives on the extremely acid (pH 4.5), infertile "cerrado" soils found in South America. For this reason, since 1974 various institutes, including CIAT (Colombia) and EMBRAPA (Brazil), have been particularly interested in this species.

Variation. From the scattered information available it appears that S. capitata is quite variable both morphologically and agronomically. Mohlenbrock (1957) noted that the specimens from Venezuela and northern Brazil were "more coarse than those from Southern Brazil and have the general vegetative appearance of S. scabra." Burt et al., (1971) mentioned that their accessions, three in number, were highly persistent but of low vigor. Grof et al. (1979) noted the occurrence of marked agronomic and morphological variation within the species.

Accessions from southern Brazil, when grown in Colombia, flowered early and produced seed yields of up to 1000 kg/ha; their vegetative yield was low. Accessions from northern Brazil behaved in the reverse fashion, i.e., producing high vegetative yield and low seed yield.

Adaptation. S. capitata (Figure 4.4) has a limited discontinuous pattern of distribution being found only in parts of Brazil and Venezuela. Within these countries it is found in a wide range of climatic conditions, from semi-arid tropical areas in Bahia, Brazil (10° to 12°S), to the almost subtropical conditions of southern Mato Grosso and Minas Gerais, Brazil (20° to 22°S), and, in one instance, to the humid tropical coast of Pernambuco (8°S) (de Lima, pers. comm.). Reference to edaphic information within these areas (Burt, Isbell, and Williams, 1979) suggests that S. capitata thrives only on the very acid, (pH 4.5), infertile "Cerrado" soils. Such observations are consistent with the results obtained from nutritional and rhizobial studies (see Sec. III).

Agronomic usage. On soils of medium acidity (pH 5.5 to 6.5), S. capitata has consistently failed to nodulate and has died. It has, however, thrived on the soils of the Llanos and Cerrado areas where its competitive ability with associate grasses, together with its tolerance of phosphorus deficient soils and its anthracnose resistance, appear to suit it to low input grazing systems. Grof (pers. comm.) has pointed out that the early and late flower-

Figure 4.4. Stylosanthes capitata.

ing forms could lend themselves to two types of management systems. Tne early, heavy seeders could suit a system of deferred
grazing[6] in which the seed could virtually be used as a protein
supplement in the dry season, whereas the late flowering, more
vegetative forms could be utilized in continuously stocked systems.

Present status of plant collections. Grof (pers. comm.) has
supplied information indicating that S. capitata collections have
recently very significantly improved. CIAT, (Colombia) collections now include material from many of the main areas in which
the species is known to occur and the EMBRAPA, Brazil collections
are also growing. This will now permit a realistic estimation of
the potential of the species. Should this potential be realized
then future plant exploration would be warranted because examination of herbarium material indicates the presence of variability
not found in the present collections (Grof, pers. comm.).

---

[6]Permitting grazing only during the initial part of the rainy season and then closing the field to the animals until the dry season
begins.

S. viscosa

This species, together with S. scabra and S. fruticosa, attracted considerable interest when it was found to persist and yield well in dry tropical climates (Burt et al., 1974); previously only the annual S. humilis could be considered for such areas. Later work (Edye et al., 1974) confirmed these results but showed that the yield of S. viscosa was generally lower than that of S. scabra. Two other factors mitigated against further efforts to domesticate this species, the anthracnose susceptibility of some accessions and several instances of unpalatability to grazing animals.

There are several reasons why further work on the species is warranted and, almost certainly, would have been carried out had the less problematical species (S. hamata, S. scabra, and S. capitata) not been available. The species is well adapted to a variety of tropical environments (Burt et al., 1974). Lack of palatability to animals has not been apparent in many conditions; indeed, unpalatability at certain times of the year may be regarded as a virtue (Gillard and Fisher, 1978; Teitzel and Middleton, Section I). From the restricted amount of information available it appears that some forms of S. viscosa are adapted to soil conditions intermediary to those of S. scabra and S. capitata (Burt, Isbell, and Williams 1979). Where S. viscosa has been tested in such situations it has outyielded S. scabra. We have examined only a small fraction of the variation present in this species. In nature S. viscosa can be found over a greater range of climatic types than any other Stylosanthes species (from humid tropical areas with precipitation greater than 2000 mm to tropical arid areas with less than 300 mm annual rainfall). Geographically it has a spectacular distributional range extending from Texas and northwest Mexico to southern Brazil and Paraguay. Our present collections, mainly from Brazil, are quite inadequate.

S. fruticosa

A full account of S. fruticosa has been presented by Skerman (1970). In the literature this species has been called S. mucronata, S. bojeri, and S. flavicans. It is an African/Indian species highly regarded as stock feed in Africa where it has several common names which, when translated, mean "rat-feed" and "sheep-feed."

S. fruticosa is a variable species in which, as in S. scabra, plant type can be related to the geographical zones and climatic conditions from which the plants were collected (Edye et al., 1974; Burt, Reid, and Williams, 1976). Agronomically it is similar to S. scabra, being a woody shrub which yields and perennates well over a range of tropical environments. Little attention has been paid to this species in more recent years because it is susceptible to anthracnose and it outcrosses readily with itself and other species (Verdcourt, 1970), rapidly producing a mixture of off-types.

S. fruticosa thrives in areas too cold or too dry for species such as S. hamata and where anthracnose is not a problem in such situations. It is resistent to both under- and overgrazing and to fire. Like S. viscosa further attempts to domesticate this species would have been forthcoming had not S. scabra and S. hamata been less problematical.

In view of its outcrossing habit and the wide range of material available to us further collections of this species do not warrant a high priority at the present time. The exception would be the material from India where the sole accession is morphologically quite distinct from the African material but has yet to be incorporated in evaluation work.

SPECIES OF LIMITED VALUE

Because of the tremendous range of naturally occurring variation available to us it is tempting, and in many instances necessary, to shelve species which are not of immediate practical use; S. viscosa and S. fruticosa are cases in point. The species considered here are perhaps even more problematical and are receiving scant attention at the present time.

S. subsericea is an annual or short-lived perennial which is similar in many regards to S. hamata. It is a poorly collected species, native to Central America. After exciting considerable interest because of its high yielding characteristics and ability to grow in cooler tropical areas (Burt, Reid, and Williams, 1976) it proved to be very susceptible to anthracnose. It has been 'shelved' in favor of S. hamata.

S. erecta is a relatively vigorous, highly persistent species native to the more humid tropical zones of Africa. It received some attention in that area (Whyte et al., 1953) but failed to attract attention in Australia because the humid tropical zones were (before the outbreak of anthracnose) well serviced by S. guianensis. Also, limited resources and lack of seed (the accessions available flowered late) resulted in its exclusion from the "new species program".

S. montevidensis and S. angustifolia were also excluded from this program: S. montevidensis because of its poor vigor and sparsely branching, stemmy growth, and S. angustifolia because of its poor persistence and moderate yield (Edye et al., 1974).

SPECIES LITTLE TESTED

We have seen how poorly known species, sometimes of very limited distribution, can rapidly assume major importance; S. capitata, agronomically unknown in 1974, can be useful for particular soil conditions, such as prevail over an area of about 2 million km^2 in South America. Therefore, we must gather what fragmentary evidence is available to indicate, albeit not very authoritatively, which species might be of significance in the future.

There are some fourteen species to consider, and we shall illustrate the procedure adopted, and the problems encountered, by reference to the species S. sympodialis. Thereafter we will briefly summarize our conclusions about the remaining species.

The potential value of S. sympodialis first became apparent to us after a series of plant geographical studies designed to select areas for future collecting missions (Burt and Reid, 1976). After nominating the types of climate most likely to contain pasture legumes suitable for use in dry tropical conditions, it was apparent that parts of coastal Ecuador and Peru and the Galapagos Islands could be of interest. A review of the literature revealed that at least one of these areas, Ecuador, also contained the types of soil (heavy clays) for which tropical pasture legumes are notably unavailable. From the scanty information available it appeared that S. sympodialis could be found in the area in which these clay soils were located. A further literature review was undertaken, this time to ascertain whether the general form of the plant might suit it to the pastoral situation. If so then the collection should be warranted. The picture to emerge, however, was far from clear. Wiggins and Porter (1971) state that "According to Mohlenbrock (1957), the collections of Stylosanthes from the Galapagos Islands, are all referable to S. sympodialis. In previous treatments such material was cited as S. scabra Vog., ... known only from the mainland, especially Brazil. S. psammophila, a variant with longer leaflets, is also known on the mainland, in Ecuador and Peru." Subsequent reviews have shown that the species mentioned are associated with different soil types, S. sympodialis with clays, S. psammophila with sandy soils, and S. sympodialis (of the Galapagos Islands) with tufaceous, coarse textured soils. They also seemed to differ in plant habit, perenniality and various other features of agronomic interest. It is difficult to allocate priorities for collection on the basis of such information.

Of the remaining species, six are tropical. S. calcicola resembles some forms of S. hamata and, like these, is found on alkaline soils, sometimes in subtropical areas. S. tuberculata is similar to S. scabra taxonomically in general habit and in the types of environment in which it is found. Mohlenbrock (1957) gives it a wide distributional range and its non-occurrence in our collections is surprising. Little is known of S. macrocarpa or S. figueroae, both of which have a very limited distributional pattern (the former near Oaxaca, Mexico and the latter near Cali, Colombia). S. cayenensis (from wet areas in French Guiana) and S. nervosa (found infrequently in Argentina, Bolivia, Peru, and Venezuela) are virtually unknown to us.

The remaining species are subtropical or, in the case of S. biflora, subtropical to temperate. This latter species is quite variable and is found exclusively in the U.S.A.

S. braceata is a perennial species which, in general appearance, is somewhat similar to S. capitata. It is, however, found in more southerly, temperate latitudes than the latter species and could be valuable in subtropical areas. S. mexicana is a variable species, some members of which generally resemble S. subsericea and other which are strong perennials. Like S. sericeiceps, noted by Mohlenbrock (1957) to resemble S. sympodalis, they could be of agronomic interest. S. leiocarpa is persistent but low yielding

(Edye et al., 1974). S. hippocampoides and S. macrosoma are similar to each other and to S. montevidensis and some forms of S.
guianensis.

Several of these species could be of interest to us because
they resemble species of known value or because they are found on
problem soils or in stringent climates.   Others, particularly
those which are rare or have only been found in restricted geographic areas, merit attention to insure that they are not lost.
Obviously, further collection is warranted.

PESTS AND DISEASES

Until very recently Stylosanthes has been noted for its relative freedom from pests and diseases; such problems as did occur
were sporadic and of limited distribution (Skerman, 1977).   Recently, however, stem boring insects have attacked some of the
woodier species (Anon., 1978) in South America and anthracnose
disease has posed a general threat.

Anthracnose of Stylosanthes was first recorded in Brazil in
1937 (Anon., 1937) and severe damage in Schofield stylo in Bolivia
was reported in 1965 (Anon. 1973).   Since then, anthracnose on
Stylosanthes has been found in Central and South America (Duron,
1969; Anon., 1973); Northern Australia in 1969 (B. Grof, pers.
comm.) and in 1973 (Pont and Irwin, 1976); Florida in 1971 (Sonoda
et al., 1974); and recently in Thailand (Lenne and Sonoda, 1978a).

With increased plantings of Stylosanthes, anthracnose has become a widespread and damaging disease, affecting all species currently under evaluation (Lenne and Sonoda, 1978a and 1978b).   It
causes considerable damage in Central and South America (Anon.,
1973, 1976, 1978; Baldion et al., 1975); causes severe losses of
seed and pasture in some regions of Northern Australia (Pont and
Irwin, 1976; O'Brien and Pont, 1977); and losses of over 50 percent dry matter in Florida (Lenne and Sonoda, 1979a).

Causal Fungi

Two pathogenic anthracnose fungi, Colletotrichum gloeosporioides and C. dematium f. truncata, have been identified.   C. gloeosporioides was identified in Bolivia in 1965 (B. Grof, pers.
comm.) and in Florida (Sonoda et al., 1974) in 1971; in 1977 both
pathogens were found in Stylosanthes seed from Australia and in
anthracnosed plants from Florida, U.S.A. (Lenne and Sonoda,
1978a).   Lenne and Sonoda (1978a, b, and 1979b) proved the pathogenicity of Australian and Florida isolates of C. dematium f.
truncata in Florida.

Both anthracnose fungi produce fruiting bodies called acer-
vuli in which masses of one-celled, hyaline spores develop, especially under moist conditions.   Cultural and morphological characters, however, readily separate the two fungi (Lenne and Sonoda,
1978a): C. dematium f. sp. truncata produces cream-colored spore
masses whereas C. gloeosporioides produces orange masses; morphologically, the curved, sickle-shaped spores and long, dark, taper-

ed setae of C. dematium f. sp. truncata are readily distinguished from the straight, cylindrical spores and short, slightly tapered setae of C. gloeosporioides.

Symptoms

On a given plant, the disease symptoms caused by both anthracnose fungi are identical (Lenne and Sonoda, 1978a) but they vary with Stylosanthes species and environmental conditions (Lenne and Sonoda, 1979b).

During warm and hot, humid conditions, elliptical to rounded lesions 0.5 to 5 mm in size with pale centers and dark margins develop on leaves, petioles, flowers, and stems. On very susceptible lines, these spots coalesce; leaves turn yellow and fall and cankers form on various parts of the plant causing leaf, flower and seed loss, stem girdling, and plant death (Sonoda, 1974; O'Brien and Pont, 1977; Irwin and Cameron, 1978; Lenne and Sonoda, 1978a and 1979b).

On S. guianensis and its varieties, elliptical to irregularly-shaped dark brown lesions of variable size develop on leaves, petioles, flowers, and stems. As these enlarge and coalesce, leaves become distorted, chlorotic, and shrivelled; leaves, flowers, and seeds drop, stems blacken and the plant dies (Lenne and Sonoda, 1979b).

Most lines of S. bracteata and S. capitata in Colombia are only slightly affected by anthracnose, which appears as dark spots or irregularly shaped superficial lesions, rarely causing leaf loss (J.M. Lenne, unpublished data).

As the weather cools, lesion margins broaden and coalesce and notable stem blackening occurs on surfaces exposed to the sun; spots and lesions blacken on S. scabra and S. guianensis (Lenne and Sonoda, 1979b).

Seed-Borne Infection

C. gloeosporioides and C. dematium f. truncata have been isolated from Stylosanthes seed harvested in Australia, Florida and South America (Ellis et al., 1976; Lenne, 1977; Lenne and Sonoda, 1978a; J.M. Lenne, unpublished data). Seed-borne anthracnose fungi are important because they can be widely and rapidly transmitted from one part of the world to another. Rapid spread of anthracnose throughout Northern Queensland, Australia (Irwin and Cameron, 1978) and introduction of anthracnose into Florida from South America (R.M. Sonoda, pers. comm.) have been attributed to seed-borne infection. Infection of Stylosanthes seed with anthracnose fungi considerably reduces germination and seedling development (Lenne and Sonoda, 1979c); affecting the establishment and persistence of stylo pastures by reducing seedling number and vigor.

Pathogen Specialization

Several strains of C. gloeosporioides (pathogenic specialization) have been found in South America (Baldion et al., 1975), Australia (Irwin and Cameron, 1978), and Florida (J.M. Lenne, unpublished data). In addition, sexual reproduction further increases the potential for production of new strains that may infect presently resistant stylos (Lenne and Sonoda, 1979b).

Pathogenic specialization between sources of Colletotrichum spp. from Australia and Florida has been noted (Lenne and Sonoda, 1978b and 1979b), heightening the possibility of introducing different strains from one country to another.

Control

The selection or breeding of pasture plants for disease resistance or tolerance is the most economical means for control. Pathogenic specialization and sexual reproduction within anthracnose fungi may make breeding for resistance difficult. Selection for resistance, preferably during field evaluation, is a better approach.

Promising Stylosanthes accessions should be evaluated under conditions representative of the most important pasture regions. Plants should be exposed to all strains of anthracnose fungi and resistant or tolerant accessions selected. Selection for resistance by field evaluation is particularly suited to South America where the existence of large numbers of indigenous anthracnose fungi (Anon., 1978) makes systematic glasshouse screening unfeasible. Resistance of S. capitata accessions to anthracnose in Colombia was largely determined from field evaluation (Grof et al., 1979).

Although resistance is most desirable, other control measures should be considered. High grazing pressures could reduce inoculum levels and produce a microclimate less suitable for anthracnose fungi, thus reducing disease. Burning could be used to control anthracnose if stylos can tolerate periodic burning. In Canada, fungal diseases of lucerne (Medicago sativa) were successfully controlled by burning (Dickson, 1956).

Use of fungicides to control pasture diseases is generally not feasible due to the large areas involved. For seed production, however, fungicides may be economically used in producing high yields of high quality seed. Benlate (benomyl) was most successful in reducing seed infection by anthracnose fungi in Florida (J.M. Lenne, unpublished data).

PLANT IMPROVEMENT PROGRAMS

At present, introduction programs are providing a wide range of adapted cultivars. As the best agronomic types are identified, future collection should be directed to the areas which have furnished such material. Only after intensive evaluation of the collected material will it be possible to assess the need for general improvement through plant breeding.

There is, however, considerable scope for genetic studies to develop a sound framework for future plant breeding. Species may be diploid (2n = 20), tetraploid (2n = 40) or hexaploid (2n = 60) (Cameron, 1967) and, within S. hamata, both diploid and tetraploid forms were present (Date et al., 1979). In the diploid species chromosome size varied from S. humilis (small), through S. hamata (medium) to S. guianensis (large) (Cameron, 1967). These differences in chromosome size were clearly seen in interspecific hybrids between S. humilis and S. guianensis and between S. humilis and diploid S. hamata (Cameron, 1968). Colchicine[7] treatment of these sterile hybrids produced partly fertile tetraploids but much more research is needed to restore seed set to a satisfactory level. The potential for interspecific hybridization at the tetraploid level is also being explored in crosses between the Verano cultivar of S. hamata and three contrasting forms of S. scabra (Cameron, unpublished). Although F_1 hybrids have low fertility, there is substantial fertility improvement in the F_2. As anticipated, there is a very diverse range of plant types in the F_2, offering considerable scope for selection. These preliminary interspecific studies are now being extended by Stace (pers. comm.) and it is hoped that these will help to relate genetic differentiation to the climatic, edaphic, and biotic adaptation of the species concerned. Intraspecific crosses among the diverse forms of S. guianensis are also being made. High levels of infertility indicate major genetic differentiation between them (Cameron, 1977).

Improvement of S. guianensis for the Subtropics

In most of the productive Stylosanthes species a large number of variable introductions has been collected, and new ecotypes can be expected from future plant exploration. The fine-stem stylo form of S. guianensis however, has a fairly restricted distribution in subtropical areas of South American and several plant collecting missions have not located any ecotypes superior to the Oxley cultivar. Fine-stem stylo is unique among Stylosanthes species for its survival through the freezing winter temperatures (-6° to -8°C) of the Queensland, Australia subtropics, but its requirements for specific nodule-forming bacteria (Date and Norris, 1979) and its low yield limits its utilization. The feasibility of transferring desirable nodulation and yield characters from tropical forms of S. guianensis to fine-stem stylo is being investigated (Cameron, 1977). Vigorous F_1 hybrids were produced from some crosses but pollen fertility was low and seed set was very poor. With this low fertility, selection pressure for the desired characters has been low but some of the F_6 lines now appear to combine the good winter survival of the fine-stem stylo parents with the nodulation characteristics and high yield of the tropical forms. Seed set has improved steadily from the F_2 generation and should be adequate for most lines following seed multiplication in the F_7 and F_8 generation.

---

[7]An alkaloid produced by the autumn-crocus (Colchicum spp. f. Liliaceae) which is used to produce chromosome doubling in plant cell nuclei.

Breeding for Resistance to Pests and Diseases

Anthracnose disease of Stylosanthes incited by the fungus Colletotrichum gloeosporioides has caused serious damage in Florida (Sonoda, 1974), South America (Baldion et al., 1975), and Australia (Pont and Irwin, 1976). The Australian cultivars of S. guianensis have been devastated in Colombia, and lines with much better resistance were found during an extensive program of plant collections and disease screening (Anon., 1975). Irwin and Cameron (1978) found two different types of anthracnose on Stylosanthes in northern Australia and subsequently Type B has devastated stands of the Schofield and Endeavour cultivars of S. guianensis on the wet tropical coast. At present, cv. Cook is fairly resistant to Type B but is severely damaged by the Colombian isolate of anthracnose (Baldion et al., 1975). Research on the disease is continuing and anthracnose resistance has now been included as a major objective in the fine-stem stylo breeding program.

In Colombia a stemborer (Caloptilia sp. B. Grof, pers. comm.) has badly damaged species with thicker woody stems, and the CIAT selection programs aim to identify genotypes with resistance to both stemborer and anthracnose. S. capitata is outstanding in this regard and is also very well adapted to the acid soils which occur over vast areas of the South American continent (Grof et al., 1979). A highly variable collection of ecotypes has been assembled and could form the basis for a breeding program once clear objectives have been defined.

Conclusions

With its wide adaptation to tropical and subtropical regions and its tolerance of infertile soils, the Stylosanthes genus has tremendous potential for improvement of grazing lands. The major task of collection and evaluation of germplasm is well under way in Australia and Colombia. A wide range of new cultivars will be developed by direct selection among introductions and will fullfill most of the immediate requirements for plant improvement. Progress in the breeding program with fine-stem stylo suggests that breeding may have a role to play in specific situations where existing genotypes are unproductive and poorly adapted. The advent of anthracnose disease and stemborer indicates that screening for pests and disease will become an integral part of the plant improvement programs. Breeding for resistance may become necessary in the future.

THE ECOLOGY OF STYLOSANTHES-BASED PASTURES

For a considerable period our thoughts about the ecology of Stylosanthes-based pastures were strongly influenced by concepts from more temperate latitudes. This is not surprising: S. humilis, at one time hailed as the "subterranean clover of the Tropics" is quite similar ecologically and agronomically to that species. The domestication of perennial species adapted to the same climates as S. humilis raises numerous ecological questions which merit some mention.

From the limited, scattered evidence available to us we suspect (but cannot prove) that pastures based on annuals will have quite different agronomic and ecological characteristics from those based on perennials. In nature the annual/short-lived species (S. hamata and S. humilis) are found in highly disturbed situations: they are 'colonizing' species or weeds. We have seen that S. humilis only achieves dominance in similar pastoral situations, i.e. when it is heavily grazed. Another 'weedy' pastoral characteristic, a 'relatively' high mineral requirement (Grof et al., 1979), is again noticeable in the pasture situation; phosphorus is necessary to insure high yields. There are both advantages and disadvantages of pastures which essentially constitute a colonizing stage. On the positive side we have a 'resilient' system which allows the cattle manager to over- or undergraze for short periods without doing permanent damage. On the negative side we need to fertilize and could have erosion problems in the dry season (most of the pasture species being annuals which are then dead).

Although we know extremely little about the perennials it would seem that they have different advantages and disadvantages. The more shrubby perennials, in particular, stay green in the dry season and once established are not susceptible to shading by tall grasses, but they are slow to establish and can be killed by overgrazing. It appears that they are adapted to situations with less grazing pressure than S. humilis. Certainly, in the area from which such perennials were collected, this seems to be the case.

Several pieces of evidence suggest that our knowledge of such species is so limited that we cannot sensibly contemplate their value. Although we now know that S. scabra can thrive on soils so infertile that fertilizer application was previously considered essential, we know nothing about the nutritional value of these plants. In preliminary studies, Playne (pers. comm.) found evidence suggesting that the phosphorus in similar plants was largely unavailable to the grazing animal. This could be desirable because the plants could provide nitrogen to the system without depleting the slender store of soil phosphorus. But it could be undesirable if the animal relies on phosphorus from the plant. Preliminary discussions suggest this could be overcome by management strategies, but this is far from proven. Ecologically such 'low phosphate' perennials seem desirable--certainly their presence has promoted the growth of more desirable grass species and invasion by broad leaved weeds has not occurred. The final outcome of such ecological change is uncertain--we have not yet investigated whether biologically fixed nitrogen can be added to such infertile soils without adding phosphorus at the same time.

Until we have a better basic understanding of such ecosystems, with the grazing animal included as part of the system, it will be difficult to appreciate or utilize the variation to be found in plant collections.

CONCLUSIONS

1. Stylosanthes is a genus of major economic significance in the tropics. The potential of the genus is, however, far from exhausted; it is still relatively easy to domesticate 'new' species and much useful variation can be found in species of established value.

2. If the potential of this genus is to be exploited, better collections or 'gene banks' are necessary. At the present time many areas and species have been disregarded; conservation has not been considered.

3. Our knowledge of these species is primarily limited to those introductions of proven agricultural value.

4. There is an urgent need to provide more meaningful descriptions of the plants involved, either through taxonomic methods or by other means, so that information can be easily exchanged.

5. There is also a need to more clearly define the niches for which improved forage legumes are required so that both existing and future collections can be rapidly evaluated. These definitions should be in terms of topography, soil texture and pH, rainfall and temperature characteristics, and present and potential land use systems.

REFERENCES

Anon. 1937. Informacoes sobre algunas plantas forrageiras. Publicacao da Seccao de Agrostologia e Alimentacao dos Animaes. No. 1. 1937.
Anon. 1973. Centro Internacional de Agricultura Tropical (CIAT), Annual Report 1972. Cali, Colombia.
Anon. 1975. Centro Internacional de Agricultura Tropical (CIAT), Annual Report 1974. Cali, Colombia.
Anon. 1976. Centro Internacional de Agricultura Tropical (CIAT), Annual Report 1975. Cali, Colombia.
Anon. 1978. Phaseolus - the CIAT collection. Plant Genetic Resources Newsletter No. 36, 21-22.
Baldion, R.W., Lozano, J.C., and Grof, B. 1975. Evaluacion de resistencia de Stylosanthes spp. a la antracnosis (Collectotrichum gloeosporioides). Fitopatologia 10: 104-108.
Blake, S.F. 1924. Stylosanthes sericeiceps. Contributions from the National Herbarium 20: 524.
Brolmann, J.B. 1978. Flood tolerance in Stylosanthes, a tropical legume. Proceedings of the Soil and Crop Science Society of Florida 37: 37-39.
Burt, R.L., Edye, L.A., Williams, W.T., Grof, B., and Nicholson, C.H.L. 1971. Numerical analysis of variation patterns in the genus Stylosanthes as an aid to plant introduction and assessment. Australian Journal of Agricultural Research 22:

176

737-757.

Burt, R.L., Williams, W.T., and Compton, J.F. 1973. Variation within naturally occurring Townsville stylo (Stylosanthes humilis) populations; changes in population structure and some agronomic implications. Australian Journal of Agricultural Research 24: 703-713.

Burt, R.L., Edye, L.A., Williams, W.T., Gillard, P., Grof, B., Page, M., Shaw, N.H., Williams, R.J., and Wilson, G.P.M. 1974. Small-sward testing of Stylosanthes in northern Australia: preliminary considerations. Australian Journal of Agricultural Research 25: 559-575.

Burt, R.L., and Reid, R. 1976. Exploration for, and utilization of, collections of tropical pasture legumes III. The distribution of various Stylosanthes species with respect to climate and phytogeographic regions. Agro-Ecosystems 2: 319-327.

Burt, R.L., and Miller, C.P. 1975. Stylosanthes - a source of pasture legumes. Tropical Grasslands 9: 117-123.

Burt, R.L., and Williams, W.T. 1975. Plant Introduction and the Stylosanthes story. Australian Meat Research Committee Review No. 25, 1-23.

Burt, R.L., Reid, R., and Williams, W.T. 1976. Exploration for and utilization of, collections of tropical pasture legumes. I. The relationship between agronomic performance and climate of origin of introduced Stylosanthes species. Agro-Ecosystems 2: 293-307.

Burt, R.L., Isbell, R.F., and Williams, W.T. 1979. Strategy of evaluation of a collection of tropical herbaceous legumes from Brazil and Venezuela I. Ecological evaluation at the point of collection. Agro-Ecosystems 5: 99-117.

Burt, R.L., and Williams, W.T. 1979. Strategy of evaluation of a collection of tropical herbaceous legumes from Brazil and Venezuela III. The use of ordination techniques in evaluation. Agro-Ecosystems 5: 135-146.

Cameron, D.F. 1967. Chromosome number and morphology of some introduced Stylosanthes species. Australian Journal of Agricultural Research 18: 375-379.

Cameron, D.F. 1968. Studies of the ecology and genetics of Townsville lucerne. Ph.D. Thesis, University of Queensland, Australia.

Cameron, D.F. 1977. Improving subtropical Stylosanthes guyanensis by hybridization with tropical forms - a progress report. In "3rd International Congress of the Society for the Advancement of Breeding Researches in Asia and Oceania (SABRAO). Plant Breeding Papers 1". Canberra, Australia 14b pp. 42-45.

Cameron, D.F. and Mannetje, L. 't 1977. Effects of photoperiod and temperature on flowering of twelve Stylosanthes species. Australian Journal of Experimental Agriculture and Animal Husbandry 17: 417-424.

Date, R.A. and Norris, D.O. 1979. Rhizobium screening of Stylosanthes species for effectiveness in nitrogen fixation. Australian Journal of Agricultural Research 30: 85-104.

Date, R.A., Burt, R.L., and Williams, W.T. 1979. Affinities between various Stylosanthes species as shown by rhizobial, soil pH and geographic relationships. Agro-Ecosystems 6: 57 67.

De Sousa Costa, N.M. and Ferreira, M.B. 1977. O Genera Stylosanthes no Estado de Minas Gerai, EPAMIG, Belo Horizonte, Brazil.

Dickson, J.G. 1956. Diseases of Field Crops. 2nd. ed. New York. McGraw-Hill. 517 pp.

Dubey, H.D., Woodburg, R., Spain, G.L., and Rodriquez, R.L. 1974. A survey of indigenous tropical legumes of Puerto Rico. Journal of Agriculture of the University of Puerto Rico 58: 87-98.

Duron, A.S. 1969. Investigaciones en leguminosas forrajeras tropicales (avance de resultados). Instituto Interamericano de Ciencias Agricolas de la OEA CT/20/587. Turrialba, Costa Rica.

Edye, L.A., Burt, R.L., Nicholson, C.H.L., Williams, R.J., and Williams, W.T. 1974. Classification of the Stylosanthes Collection 1928-1969. C.S.I.R.O. Division of Tropical Agronomy Technical Paper no. 15.

Edye, L.A., and Cameron, D.F. 1975. Comparison of Brazilian and naturalized Australian ecotypes of Stylosanthes humilis in the dry tropics of Queensland. Australian Journal of Experimental Agriculture and Animal Husbandry 15: 80-87.

Edye, L.A., Field, J.B., and Cameron, D.F. 1975. Comparison of some Stylosanthes species in the dry tropics of Queensland. Australian Journal of Experimental Agriculture and Animal Husbandry 15: 655-662.

Edye, L.A., Field, J.B., Bishop, H.G., Hall, R.L., Prinsen, J.H., and Walker, B. 1976. Comparison of some Stylosanthes species at three sites in central Queensland. Australian Journal of Experimental Agriculture and Animal Husbandry 16: 715-721.

Edye, L.A., Williams, W.T., Bishop, H.G., Burt, R.L., Cook, B.G., Hall, R.L., Miller, C.P., Page, M.C., Prinsen, J.H., Stillman, S.L., and Winter, W.H. 1976. Sward tests of some Stylosanthes guyanensis accessions in tropical and subtropical enviroments. Australian Journal of Agricultural Research 27: 637-647.

Ellis, M.A., Ferguson, J.E., Grof, B., and Sinclair, J.B. 1976. Transmission of Colletotrichum gloeosporioides and effect of sulfuric acid scarification on internally-borne fungi in seeds of Stylosanthes spp. Plant Disease Reporter 60: 844-846.

Gardener, C.J. 1975. Mechanisms regulating germination in seeds of Stylosanthes. Australian Journal of Agricultural Research 26: 281-294.

Gardener, C.J. 1978. Seedling growth characteristics of Stylosanthes. Australian Journal of Agricultural Research 29: 803-813.

Giacometti, D.C. 1979. The national programme of Brazil. Plant Genetic Resources Newsletter No. 36, 6-7.

Gillard, P., and fisher, M.J. 1978. The ecology of Townsville stylo-based pastures in northern Australia. In "Plant Relations in Pastures" (ed.) J.R. Wilson, CSIRO, Melbourne, Australia. pp. 340-352.

Grof, B., Schultze-Kraft, R., and Muller, F. 1979. Stylosanthes capitata Vog., some agronomic attributes and resistance to anthracnose (Colletotrichum gloeosporioides Penz). Tropical Grasslands 13: 28-37.

Hartley, W. 1949. Plant collecting expedition to sub-tropical South America, 1947-48. CSIRO Australia, Division of Plant Industry, Divisional Report, No. 7.

Hassler, E. 1919. Ex herbario Hassleriano. Novitates Paraguariensis XXIII. Fedde Repertorium, Species Novae 16: 220-224.

Holm, A. Mc R. 1973. The effect of high temperature pretreatments on germination of Townsville stylo seed material. Australian Journal of Experimental Agriculture and Animal Husbandry 13: 190-192.

Humphreys, L.R. 1967. Townsville lucerne: history and prospect. Journal of the Australian Institute of Agricultural Science 33: 3-13.

Hutchinson, J. 1964. The Genera of Flowering Plants (Angiospermae). Volume I. Dicotyledones. Oxford University Press, Oxford.

Irwin, J.A.G. and Cameron, D.F. 1978. Two diseases in Stylosanthes spp. caused by Colletotrichum gloeosporioides in Australia, and pathogenic specialization within one of the causal organisms. Australian Journal of Agricultural Research 29: 305-317.

Jones, R.K. 1974. A study of the phosphorus responses of a wide range of accessions from the genus Stylosanthes. Australian Journal of Agricultural Research 25: 847-862.

Kretschmer, A.E. 1968. Stylosanthes humilis, a summer-growing, self-regenerating annual legume for use in Florida pastures. Florida Agricultural Experimental Stations. Circular S-184.

Leeuw, P.N. de 1974. The establishment of Stylosanthes humilis in Nigerian savannah. Proceedings of the 12th International Grassland Congress, Moscow, USSR. pp. 182-193.

Lenne, J.M., and Parbery, D.G. 1976. Glomerella cingulata on Stylosanthes spp. in the Northern Territory. Australian Plant Pathology Society Newsletter 5: 24-25.

Lenne, J.M. 1977. Colletotrichum dematium on Stylosanthes spp. Australian Plant Pathology Society Newsletter 6: 53-54.

Lenne, J.M., and Sonoda, R.M. 1978a. Occurrence of Colletotrichum dematium f. sp. truncata on Stylosanthes spp. Plant Disease Reporter 62: 641-644.

Lenne, J.M., and Sonoda, R.M. 1978b. Colletotrichum spp. on tropical forage legumes. Plant Disease Reporter 62: 813-817.

Lenne, J.M., and Sonoda, R.M. 1979a. Effect of Colletotrichum gloeosporioides on yield of Stylosanthes hamata. Abstracts of Papers IX International Congress of Plant Protection and 71st Annual Meeting of the American Phytopathologies Society. Washington, D.C., U.S.A. (Abst. No. 858).

Lenne, J.M., and Sonoda, R.M. 1979b. The occurrence of Colletotrichum spp. on Stylosanthes spp. in Florida and the pathogenicity of Florida and Australian isolates to Stylosanthes

spp. Tropical Grasslands 13: 98-105.
Lenne, J.M., and Sonoda, R.M. 1979c. The effect of seed inoculation with Colletotrichum spp. on emergence, survival and
seedling growth of Stylosanthes hamata. Tropical Grasslands
13: 106-109.
McIvor, J.G. 1976a. Germination characteristics of seven Stylo-
santhes species. Australian Journal of Experimental Agriculture and Animal Husbandry 16: 723-728.
McIvor, J.G. 1976b. The effect of waterlogging on the growth of
Stylosanthes guyanensis. Tropical Grasslands 10: 173-178.
McTaggart, A. 1937. Stylosanthes. Journal of the Council of Scientific and Industrial Research 10: 201-203.
Mannetje, L. 't 1969. Rhizobium affinities and phenetic relation-
ships within the genus Stylosanthes. Australian Journal of
Botany 17: 553-564.
Mannetje, L. 't 1977. A revision of varieties of Stylosanthes
guianensis (Aubl.) SW. Australian Journal of Botany 25: 347-
362.
Mannetje, L. 't, O'Connor, K.F., and Burt, R.L. 1979. The use and
adaptation of pasture and fodder legumes. In "Advances in
Legume Science" (ed.) R.J. Summerfield. Royal Botanic Gar-
dens, Kew, England.
Mohlenbrock, R.H. 1957. A revision of the genus Stylosanthes.
Annals of the Missouri Botanical Gardens 44: 299-355.
Norman, M.J.T. and Arndt, W. 1959. Performance of beef cattle on
native and sown pasture at Katherine, N.T. CSIRO Australia,
Division of Land Use and Regional Survey, Technical Paper No.
4.
Norman, M.J.T. 1966. Katherine research station 1956-64: a re-
view of published work. CSIRO Australia, Division of Land
Research, Technical Paper No. 28.
O'Brien, R.G., and Pont, W. 1977. Diseases of Stylosanthes in
Queensland. Queensland Agricultural Journal 103: 126-128.
Pittier, H. 1944. Leguminosas de Venezuela 1. Papilionacea sp.,
Boletin Tecnico No. 5. Ministerio de Agricultura y Cria,
Servicio Botanico, Caracas, Venezuela.
Pont, W., and Irwin, J.A.G. 1976. Colletotrichum leaf spot and
stem canker of Stylosanthes spp. in Queensland. Australian
Plant Pathology Society, Newsletter 5: No. 1 . (supplement).
Robinson, P.J. and Megarrity, R.G. 1975. Characterization of
Stylosanthes introductions by using seed protein patterns.
Australian Journal of Agricultural Research 26: 467-479.
Rotar, P., Dworak, S.H., Evans, D.O. and Walker, J.L. 1981. A
selected bibliography of the pasture legumes Centrosema, Des-
modium, and Stylosanthes and other tropical and subtropical
pasture legumes. University of Hawaii, Hawaii Institute of
Tropical Agriculture and Human Resources. Research Series
002.
Schofield, J.L. 1941. Introduced legumes in North Queensland.
Queensland Agricultural Journal 56: 378-388.
Schoonhover, H.C. and Humphreys, L.R. 1974. Seed production of
Stylosanthes humilis as influenced by origin and temperature
during flowering. Crop Science 14: 468-471.

Shaw, N.H. 1961. Increased beef production from Townsville lucerne (Stylosanthes sundaica Taub.) in speargrass pastures of central coastal Queensland. Australian Journal of Experimental Agriculture and Animal Husbandry 1: 73-80.

Shaw, N.H. and Mannetje, L. 't 1970. Studies on speargrass pasture in central coastal Queensland - the effect of fertilizer, stocking rate and oversowing with Stylosanthes humilis on beef production and botanical composition. Tropical Grasslands 4: 43-56.

Skerman, P.J. 1970. Stylosanthes mucronata Willd., an important natural perennial legume in eastern Africa. Proceedings of the XI International Grassland Congress, Surfers Paradise, Australia. (ed.) M.J.T. Norman. University of Queensland Press. pp. 196-198.

Skerman, P.J. 1977. Tropical Forage Legumes. F.A.O. Plant Production and Protection Series No. 2. F.A.O., Rome.

Sonoda, R.M. 1973. Incidence of Colletotrichum leaf spot and stem canker on introductions and selections of Stylosanthes humilis. Plant Disease Reporter 57: 747-749.

Sonoda, R.M. 1974. Diseases of Stylosanthes spp. Ft. Pierce Agricultural Research Center Research Report RL-1974-4. 3pp.

Sonoda, R.M., Kretschmer, A.E. Jr., and Brohlmann, J.B. 1974. Colletotrichum leaf spot and stem canker on Stylosanthes spp. in Florida. Tropical Agriculture (Trinidad) 51: 75-79.

Standley, P.C. and Steyermark, J.A. 1946. Flora of Guatemala. Leguminosae. Fieldiana : Botany 24: pp. 350.

Stehle, H. 1956. Survey of forage crops in the Caribbean. Caribbean Commission Central Secretariat, Port-of-Spain, Trinidad.

Stonard, P. and Bisset, W.J. 1970. Fine-Stem Stylo: a perennial legume for the improvement of subtropical pasture in Queensland. Proceedings of the XI International Grassland Congress, Surfers Paradise, Australia (ed.) M.J.T. Norman. University of Queensland Press. pp. 153-158.

Swartz, O. 1788. Nova genera et species plantarum seu prodromus descriptionum vegetabilium, maximam partem incognitorum quae subitinere in Indiam occidentalem annis 1783-87 digesset Olof Swartz. Holmiae, M. Swederi.

Sweeney, F.C. and Hopkinson, J.M. 1975. Vegetative growth of nineteen tropical and sub-tropical pasture grasses and legumes in relation to temperature. Tropical Grasslands 9: 209-217.

Talineau, J.C. 1970. Influence of climatic factors on forage production in the Ivory Coast. Cahiels ORSTROM, Biologie, No. 14: 51-76 (French).

Taubert, P. 1891. Monographie der Gattuay Stylosanthes. Verbandlunger des Botanischer Vereins der Provinz Brandenburg 32: 1-34.

Tuley, P. 1968. Stylosanthes gracilis. Herbage Abstracts 38: 87-94.

Verdcourt, H. 1970. Studies in the Leguminosae-Papilionoideae for the "Flora of Tropical East Africa" I. Kew Bulletin 24(2): 1-70.

Vogel, J.R.T. 1838. Stylosanthes. Linnaea 12: 63-71.

Whyte, R.O., Nalsson-Leissner, G., and Trumble, H.C. 1953. Legumes in Agriculture. FAO Agricultural Studies No. 21, Rome, Italy.

Wiggens, I.L. and Porter, D.M. 1971. Flora of the Galapagos Islands. Stanford University Press, Stanford, California. pp. 635.

5
Rhizobium Germplasm Resources for *Centrosema, Desmodium,* and *Stylosanthes*

R. A. Date

The majority of isolates of Rhizobium for the genera Centrosema, Desmodium, and Stylosanthes is held in four collections: 1) The Division of Tropical Crops and Pastures, CSIRO, Brisbane, Australia; 2) The Tropical Pastures Program, CIAT, Cali, Colombia; 3) Biological Nitrogen Fixation Program, EMBRAPA, Seropedica, Brail; 4) Niftal, University of Hawaii, Paia, Hawaii, U.S.A. Isolates from Stylosanthes number approximately 2,000 with a further 600 to 800 each for Centrosema and Desmodium.

Information on symbiotic performance (effectiveness on a range of accessions) is limited to very few cultures. For example, in the CSIRO collection less than 40 strains for Stylosanthes have been studied in detail. Despite the relatively large number of isolates, the range of rhizobial germplasm is limited. For instance, most of the isolates for Stylosanthes originate from S. guianensis, S. hamata, S. humilis, and S. scabra. There are very few isolates from species such as S. calcicola, S. capitata, S. leiocarpa, S. montevidensis, S. sympodialis, and S. viscosa even though some of these could well be of agronomic interest. If indeed there is a need to suit the rhizobia to the edaphic conditions most suited to the host, then the deficiency in suitable strains for this latter group may prevent their agronomic exploitation. For example, S. capitata and S. viscosa come from very acid soil conditions, whereas S. sympodialis is from heavy clay alkaline soils. It is also worth noting that existing collections have come primarily from Colombia, Brasil and Venezuela and to a lesser extent, the Caribbean, Costa Rica, Mexico, and· Panama. Isolates from Africa and Asia are rare even though strains such as CB756, which is effective on many S. fruticosa, S. guianensis, S. humilis, S. scabra, and S. viscosa lines, does originate from a collection in East Africa.

To a large extent, isolates in the above collections are from areas which are accessible or have active research institutes and not from areas stipulated by a deliberate collecting program.

Isolates of Rhizobium for Centrosema and Desmodium are even more limited than for Stylosanthes. Most are from C. pubescens and D. intortum, although the CIAT and CSIRO collections now cover a small range of isolates for C. plumieri, C. virginianum, D. ad-

183

scendens, <u>D</u>. <u>canum</u>, <u>D</u>. <u>distortum</u>, <u>D</u>. <u>heterocarpum</u>, <u>D</u>. <u>heterophyl</u>-<u>lum</u>, and <u>D</u>. <u>uncinatum</u>. The species and geographic representation for these two genera is very poor but follows the same distribution as that for <u>Stylosanthes</u>. Clearly, our collections of tested and proven rhizobial strains do not match the corresponding plant resource material. This deficiency is so extensive that there are many otherwise agronomically desirable plants for which the <u>Rhizo</u>-<u>bium</u> requirements are either unknown or cannot be satisfied by existing strains. There is a need to collect the appropriate rhizobia from the areas which have yielded such plant material. It is also imperative that this situation should be avoided in the future. Although the simultaneous collection of nodules and plant material is not always easy (since the optimum time for the collection of nodules does not always correspond with that for seed collection) plant collecting missions should attempt to do so. Details for the collection of nodules for the isolation of rhizobia and of the recording of site information have been described by Date and Halliday, (1979).

REFERENCES

Date, R.A. and Halliday, J. 1979. Collection of strains of <u>Rhi</u>-<u>zobium</u>. In "Handbook for the Collection, Preservation and Evaluation of Tropical Forage, Germplasm Resources" (ed.) G.O. Mott. Workshop CIAT, Colombia, April 1978. Chapter V. pp. 21-26.

6
General Conclusions

Each of these three reviews was prepared independently of the other; each involved two or more authors with different scientific interests and training. It is quite remarkable to find, therefore, that the stories they present, and the conclusions reached and suggestions made, are much in accord. Clearly, these merit attention when research priorities of the future are to be considered.

In each of the genera the general story was the same. Early, scattered introductions were tested and found to be agronomically useful. These were made commercially available and subsequently distributed and used throughout much of the tropical world. Virtually all of the detailed information available to us concerns these species but, even here, there is missing information.

The commercial use of these cultivars revealed limitations: they were susceptible to disease or simply not adapted to areas for which legumes were required. Further introduction programs rapidly unearthed more desirable genotypes, either within the species concerned or in 'new' species. The authors conclude that plant collection has been too limited in the past; there is a wealth of naturally occurring variation which is likely to be extremely useful. In two of the reviews the authors believe that plant breeding, genetic, and taxonomic studies cannot sensibly be carried out, except on a restricted scale, until representative collections are available.

At the present time, the lack of meaningful taxonomic treatments imposes severe constraints on plant collecting, introduction, and evaluation programs. There is considerable confusion between species, and individual species may be so "wide" that they contain many agronomic types. This causes tremendous problems when attempting to exchange seed or information and, as a result, it is often impossible to sensibly use data from other sources. One author suggests that the value of testing data would be greatly enhanced if climatic, soil, and nutrition data were published with the plant information; this will be discussed in the next section.

Taxonomy is, however, only one of the disciplines which we consider to be important; much more needs to be done in other fields such as microbiology and plant nutrition.

Section III
Special Considerations in Utilizing Tropical Legumes in Pastures

1
Comparative Edaphic Requirements

P. J. Robinson

The development of sown pastures in the tropics is a relatively recent phenomenon; even the most likely source of the legumes to be used was in doubt until recently, (W. Davies, 1933; J.G. Davies, 1965) and attempts were often made to use the more familiar temperate species. These performed poorly and it was recognized that the infertile, often acid soils of the tropics may have been at least partially responsible. Soil ameliorative techniques, addition of lime to correct the acidity and superphosphate to provide nutrients, proved to be expensive and often transitory.

Introduced tropical legumes often survived, however, and pioneering work helped to explain why. Andrew and Norris (1961) showed that many tropical legumes prefer acid conditions and thrive on relatively low soil mineral levels. Norris (1958, 1965) demonstrated that the rhizobia associated with such plants also prefer acid soils and form effective plant/rhizobial associations under infertile conditions.

Tropical soils, of course, vary greatly in their fertility levels. It soon became apparent that different fertility levels demanded different legumes; <u>Stylosanthes</u> became accepted as the genus for low fertility and genera such as <u>Centrosema</u>, <u>Desmodium</u>, and <u>Macroptilium</u> for somewhat higher levels (see Sections I-7, II-2, II-3, and II-4). A plethora of field and pot experiments would seem to support these conclusions. Most, however, have been carried out on a few common cultivars and their applicability to generic differences might be questioned.

To explore these nutrient requirements further and to aid in the production of mineral requirements under field conditions, Andrew and co-workers (see review by Andrew and Jones, 1978; Smith, 1978), have studied a variety of tropical legumes in some detail. The concept of the "critical value" was developed: plants are grown under different levels of a given nutrient, and their yields recorded at a given age; yield is then plotted against the concentration of the nutrient in the plant, and that concentration which allows 90 percent of maximum growth is deemed to be the "critical value". The results obtained corresponded reasonably well with field conditions; S. <u>humilis</u> had a critical value, for P, of 0.17 percent, whereas the species for more fer-

189

tile situations (Centrosema pubescens, Desmodium intortum, and Macroptilium atropurpureum) had equivalent values of 0.16, 0.21, and 0.24 percent, respectively.

Application of this method as a diagnostic tool in perennial pasture systems is, however, somewhat limited. Critical values change with age and are dependent upon the part of the plant sampled. Probably this is a result of the different translocation rates of different nutrients or different physiological rates for the same nutrient in different genera. There is also a problem in sensibly fitting the curves relating plant weight and nutrient concentration within the plant (Smith, 1978). In one published example the soil critical P value varied between 3 and 8 ppm depending upon the method used.

The ability of plants to tolerate, or thrive in, soils of low fertility has become of increasing interest as fertilizer costs have escalated. The actual mechanisms involved in the differential tolerance of plants are, however, difficult to isolate and few relevant studies have been carried out. The general belief that Stylosanthes can survive in situations which other genera cannot tolerate seems to stem from field observations and is substantiated with evidence from only one detailed study. In this, Andrew (1966) found that absorption of phosphorus by root segments of S. humilis (Townsville stylo) greatly exceeded that of Medicago sativa (lucerne or alfalfa), Macroptilium lathyroides (phasey bean), Desmodium uncinatum (Silverleaf desmodium), and barley (see Figure 1.1). The same author, however, points out the difficulties inherent in extrapolating such results to whole plants or to field situations (Andrew and Jones, 1978). Indeed, it is difficult to even define what is meant by a "phosphorus efficient" species and Blair and Cordera (1978) felt it necessary to consider four criteria: (1) the ability to produce top dry weight with a given amount of applied P, (2) top dry matter produced per unit of P taken up, (3) top dry matter produced at a constant plant P level and (4) phosphorus uptake per unit root weight. They found that the order of "efficiency" of the three species varied according to the definition used and in some instances to the time of harvest.

The latter results were obtained in nutrient and soil culture under controlled conditions. In the field the situation is even more complex and such factors as differences in root distribution between species need to be considered. Perenniality poses additional problems associated with differential translocation to the roots. The success of a perennial legume in infertile conditions is thus dependent upon a multitude of interacting factors, and this makes it difficult to develop meaningful 'standard' approaches in mineral nutrition studies. If, of course, more detailed studies are carried out then the number of species which can be studied is greatly reduced.

It is for this reason that virtually all of our knowledge on tropical legumes is restricted to "standard" plants such as the cultivars of Centrosema pubescens, Macroptilium lathyroides (formerly known as Phaseolus lathyroides), and Macroptilium atropurpureum, etc. This knowledge has, moreover, usually been gained

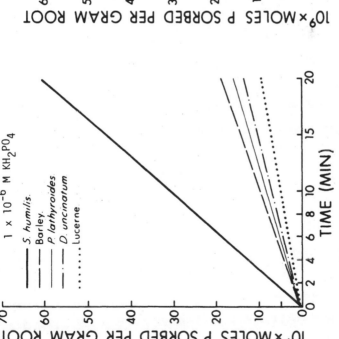

Figure 1.1.   Relationship of phosphorus uptake to time in excised roots of four legumes and barley. (Data from Andrew, 1966).

after the plants have proved to be relatively successful; they may, or may not be, representative of the species not to mention the genus (see also Section II). The work of Jones (1974) suggests that the latter is at least sometimes the case (Figure 1.2). Accessions 2 and 3, of S. hamata and S. humilis respectively, have both been successful on acid soils in Australia; these response patterns are typical of plants not requiring a high calcium environment. Accessions 1, 4, and 5, of the same species, have performed poorly on the same soils; they have response patterns typical of plants requiring a high calcium environment. Not surprisingly, on alkaline soils, their position is reversed.

These data are of interest from another point of view. Examination of the edaphic origins of the plants reveals that accessions 2 and 3 are from acid soils and that the others, where data are given, are from alkaline conditions. Teitzel and Middleton (Section I-7) described a similar situation where they showed that the fertilizer requirements for the establishment of various species could best be deduced from a knowledge of the species concerned, the native vegetation and the geological origin of the soils. Possibly knowledge of the soils from the collection areas could be used to broadly indicate the nutritional requirements of the various species. Attempts to do this have been presented by Burt et al. (1979), who examined the variation in soils from which various legumes were collected and related this to the species found. The physical nature of the soils was important and Stylosanthes, Centrosema, and Macroptilium were confined to the sandy textured surface soils. Within these major soil groups, however, the balance of the various minerals was important (Figure 1.3). On the highly acid, low P situations (quadrant A) only the genus Stylosanthes was encountered. As the phosphorus level increased, and the exchangeable K level dropped, Centrosema and Macroptilium became more prevalent. These findings are in accord with the general belief that Stylosanthes is most tolerant to conditions of low fertility and Centrosema and Macroptilium less so. Note also that a variety of species were concerned. Even more remarkable, however, is the close agreement with the work of Hall (1975). This author has shown that, when grown in competition with a grass, it is the addition of K, not P, which favors the legume in infertile conditions. The addition of P favors the grass more.

Although relatively scanty, the data in Figure 1.3 also suggest that the various Stylosanthes species differ in their fertility requirements. On extremely acid, phosphorus deficient soils, only S. capitata was found; S. viscosa and S. scabra occurred on somewhat less acid soils and S. humilis and S. guianensis in even more neutral conditions. Observations of plants from which it was not possible to collect seed, and which for various reasons did not appear in Figure 1.3, tend to support these distinctions. This suggests, therefore, that species may differ in their ability to tolerate such conditions. Current evaluation work in Australia (Burt, unpublished data) tends to confirm these findings. The work of Grof et al. (1979), however, suggests that this species ranking reflects not only tolerance but preference; S. capitata, for instance reached maximum growth levels at very low P levels,

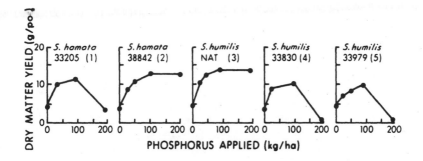

Figure 1.2. The phosphorus response of accessions of Stylosanthes hamata and S. humilis (Jones, 1974).

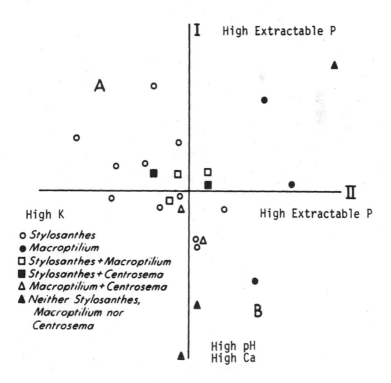

Figure 1.3. Principal coordinate analysis of 24 surface soils with the predominant genera superimposed on the sites.

<u>S. viscosa</u> at somewhat higher levels. Date (Section II-5), more-over, has shown that the associated rhizobia may be similarly adapted. Attempts to refine this approach, by which plant collecting data are used to indicate likely mineral response patterns, are currently being made (Burt, Pengelly, and Williams, 1980). It is hoped that these will be useful in selecting fruitful areas of research for those involved with mineral nutrition studies.

Other approaches are currently being developed. Bowling and Dunlop (1978) have calculated the energy expended in accumulating phosphate; they suggest that such values may be useful as "phosphate efficiency indices." Many questions, however, remain unresolved. How do we "screen" large numbers of plants for their "adaptability" to edaphic constraints? Undoubtedly, an answer would considerably expedite the provision of suitable legumes; traditional plant analyses, biochemical, and physiological techniques all appear to have limitations (Smith, 1978). Do "low fertility" species utilize the same phosphorus pool as other species or are they able to tap resources unavailable to "high fertility" species? If extraction and removal of phosphate exceeds the recirculatory input then will pastures based on such species invariably decline in productivity? Possibly most important of all, what is the overall role of "high" and "low" fertility species and what are the scientific and economic constraints which govern their long term use?

This review has been brief and, to a very large extent, uses only examples of phosphorus nutrition to illustrate the points raised. It would have been equally possible to use, for instance, trace elements, in which deficiencies are fairly common, where differential tolerances of various species occur and where there are limitations in the methods used to diagnose deficiencies (see review by Bruce, 1978). Our main thesis is that, although we have learned much about the mineral nutrition of tropical legumes over the last decade, many challenging issues remain unanswered. If biological nitrogen fixation is to assume a full, important role in tropical pasture development then solutions to these problems should be sought.

REFERENCES

Andrew, C.S. 1966. Kinetic study of phosphate absorption by excised roots of <u>Stylosanthes</u> <u>humilis</u>, <u>Phaseolus</u> <u>lathyroides</u>, <u>Desmodium</u> <u>uncinatum</u>, <u>Medicago</u> <u>sativa</u>, and <u>Hordeum</u> <u>vulgare</u>. <u>Australian Journal of Agricultural Research</u> 17: 611-624.

Andrew, C.S. and Jones, R.K. 1978. The phosphorus nutrition of tropical forage legumes. In "Mineral Nutrition of Legumes in Tropical and Subtropical Soils." (eds.) C.S. Andrew and E.J. Kamprath, CSIRO Melbourne, Australia.

Andrew, C.S. and Norris, D.O. 1961. Comparative responses to calcium of five tropical and four temperate pasture legume species. <u>Australian Journal of Agricultural Research</u> 12: 40-55.

Blair, C.J. and Cordera, S. 1978. The phosphorus efficiency of three annual legumes. <u>Plant and Soil</u> 50: 387-398.

Bowling, D.J.F. and Dunlop, J. 1978. Uptake of phosphate by white clover. I. Evidence for an electrogenic phosphate pump. Journal of Experimental Botany 29: 1139-1146.

Bruce, R.C. 1978. A review of the trace element nutrition of tropical pasture legumes in Northern Australia. Tropical Grasslands 12: 170-183.

Burt, R.L., Isbell, R.F., and Williams, W.T. 1979. Strategy of evaluation of a collection of tropical herbaceous legumes from Brazil and Venezuela. I. Ecological evaluation at the point of collection. Agroecosystems 5: 99-117.

Burt, R.L., Pengelly, B.C., and Williams, W.T. 1980. Network analysis of genetic resources data. III. The elucidation of plant/soil/climate relationships. Agroecosystems 6: 119-127.

Davies, W. 1933. The grasslands of Australia and some of their problems. CSIRO Australia. Pamph. 39.

Davies, J.G. 1965. Pasture improvement in the tropics. Proceedings 9th International Grassland Congress I. 217-220.

Grof, B., Schultze-Kraft, R., and Muller, F. 1979. Stylosanthes capitata Vog., some agronomic attributes, and resistance to anthracnose (Colletotrichum gloeosporoides Penz). Tropical Grasslands 13: 28-37.

Hall, R.L. 1975. Nutrient ecology-competition for nutrients in pastures mixes. In CSIRO Div. Trop. Agronomy Annual Report pp. 77.

Jones, R.K. 1974. A study of the phosphorus responses of a wide range of accessions from the genus Stylosanthes. Australian Journal of Agricultural Research 25: 847-862.

Norris, D.O. 1958. Lime in relation to nodulation of tropical legumes. In "Nutrition of the Legumes" (ed.) E.G. Hallsworth Butterworths; London. pp. 164-182.

Norris, D.O. 1965. Acid production by Rhizobium - a unifying concept. Plant and Soil 22: 143-166.

Smith, F.W. 1978. Role of plant chemistry in the diagnosis of nutrient disorders in tropical legumes. In "Mineral Nutrition of Legumes in Tropical and Subtropical Soils." (eds.) C.S. Andrew and E.J. Kamprath, CSIRO Melbourne, Australia. pp. 329-346.

2
Edaphic Factors

Y. Kanehiro, J. L. Walker,
and M. Asghar

INTRODUCTION

To maximize yield, the agriculturalist attempts to find, or to develop by breeding, a plant fully adapted to the environment in which it is to be grown. That environment includes the soil from which the plant draws essential nutrients and water, and the climate which can determine rate and amount of growth. More specifically, the solar energy input, the ambient temperature, and the incidence of rainfall and wind are the climatic components. Growth may be limited by departures from the norm; the capability of the plant to recover from climatic stresses is a significant aspect of adaptation.

Complicated as this may be for a single crop plant, the complexity of the ecology of a pasture is much greater. If a legume and grass are grown in association, the objective is not as simple as when selecting a species for a mono-culture. Maximum dry matter yield at a defined harvest date may be far less important than continued productivity, maintenance of the same botanical composition, and stability under whatever grazing regime is established. The protein content of the mixed herbage, its palatability, and digestibility for maximum gain in animal weight are perhaps of more consequence in assessing the quality of a pasture than total dry matter yield.

To maximize the yield of a short season crop, one can select that period of the year during which edaphic factors are optimal or near optimal. The situation with a grass-legume pasture is entirely different. In the selection of species to be used, there is clearly a compromise that must be reached. The ability of grass/legume components to survive adverse climatic factors, such as a rainless, hot period or a cloudy wet period has to be considered along with their ability to thrive when climate factors are benign.

Soil conditions are, of course, of paramount importance--the the soil as supplier of the nutrient elements essential for plant growth, and the medium from which the water requirements are met. Once again the situation with a grass-legume pasture differs from that with a short-season crop. Not only must there be adequate essential elements and absence of deleterious physico-chemical conditions, but there must be a continuity of supply if the pasture is to maintain productivity. Tropical soils are frequently deficient in one or more essential nutrient elements or become deficient under continued use despite some recyling that occurs under grazing animals.

SOIL ACIDITY AND RELATED EFFECTS

Soil pH

Acid soils frequently limit nitrogen fixation in highly weathered soils of the humid tropics. The soil pH of many Ultisols and Oxisols is below 5.0. Some Inceptisols, Alfisols, and Histosols also limit N fixation because of high acidity, particularly those in high rainfall areas. The detrimental effects of acidity may be caused by a number of factors: low pH, excessive levels of Al and Mn and deficiency of Ca and/or Mo. In many situations, it is difficult to sort out the factor of soil acidity which affects nitrogen fixation. Andrew (1978) reviewed early work that indicated that the effect of low pH (4.5 to 5.5) on tropical legume growth was not inhibitory in a healthy plant environment. Recent studies however, revealed that pH does play an important role in the legume-rhizobia system. For example, within the tropical group of legumes, individual species and cultivars within species react differently to pH. (Andrew, 1976 and Munns et al. 1977). It is sufficient to say that problems which occur with soil acidity that call for corrective measures, introducing the topic of liming tropical soils. Here again, the problem is multifaceted. Lime can be added to soil to: (a) correct pH per se, (b) add calcium (or magnesium if dolomitic lime is used), (c) decrease excessive Al and/or Mn, or (d) increase the availability of Mo.

Lime can be applied by pelleting the seed or by direct application to the soil. Most studies show a higher lime requirement for growth and nodulation than can be provided by a pellet. For example, Dradu (1974) in Uganda reported that increasing lime rate up to 6250 kg/ha gave increased dry matter yield and uptake of Desmodium intortum.

Enough work has been done in the past two decades to show that lime cannot be applied to soils in the tropics according to standards for soils in the temperate zone. It is generally accepted that liming above pH 6.0 is unwise because of the possible creation of deficiencies in micronutrients and P. Furthermore, some tropical legume species may be adapted to low pH and Ca; therefore, raising the soil pH to levels much above 6.0 can cause deleterious effects (Munns, 1976). This point may be academic, because large amounts of lime are often required to increase soil pH beyond pH 6. The classical way to determine how much lime to apply to raise soil pH is to work with a buffer curve prepared by plotting the pH as increments of a standard base added to a hydrogen-saturated soil. Another method, and the preferred one for tropical soils, is based on exchangeable aluminum data. This method will be discussed later in the subsection dealing with aluminum.

Calcium

A low exchangeable Ca supply in soil is usually associated with high soil acidity, although there are exceptions to this rule. Tropical legumes generally tolerate low Ca better than temperate legumes. In an early work, Andrew and Norris (1961) compared yield performances of three temperate legumes (Medicago truncatula, Medicago sativa, and Trifolium repens) with those of five tropical legumes (Desmodium uncinatum, Indigofera spicata, Centrosema pubescens, Stylosanthes bojeri, and Macroptilium lathyroides) after adding varying rates of CaCO₃ to a Ca-deficient soil. They reported that lime increased yields of the Medicago and Trifolium more than others. Calcium percentages in tops of the two groups of legumes did not differ. They concluded that the tropical species have a superior ability to absorb Ca from the soil. Later studies have shown that Ca requirements differ among species, and other soil factors, such as high manganese (Dobereiner and Aronovich, 1965) and availability of molybdenum (Mears and Barkus, 1970), affect the need for lime/Ca.

With high Ca levels in the soil Andrew (1976) found adverse effects caused by 2mM Ca concentration in Stylosanthes humilis and Lotononis bainesii. Such effects by liming were similarly noted by Munns et al. (1977) for the above two species and by Souto and Dobereiner (1969) for species of Stylosanthes.

It is difficult to separate the individual effects of pH and Ca. Andrew (1978) has shown that the effects of low and high pH dominate those of Ca concentration, and that the degree of pH effect varies between species. The largest effect of Ca occurs at the intermediate pH range. Lee and Wilson (1972) conducted a pot experiment with Glycine wightii on a soil low in Ca and pH 5.3 and found that it responded to both Ca and pH. The pH effect did not involve Mo supply or the alleviation of Mn and Al toxicities.

Trigoso and Fassbender (1973) reported results of a pot experiment with an andesitic soil (Ultisol) from Costa Rica. Dry matter production of Macroptilium atropurpureum, Desmodium intortum, Glycine javanica, and Centrosema pubescens increased quadratically with lime application. Highest yields were obtained with 6-10 me Ca + Mg/100 g soil. Correction of Ca deficiency alone requires much less lime than changing soil pH (Loneragan, 1973; Munns, 1976; Pearson, 1975). Sanchez and Isbell (1979) reported that 0.1 to 0.5 t/ha of dolomitic lime corrected Ca and Mg deficiencies in Oxisols and Ultisols of South America. Calcium salts, such as gypsum, can also be used as a Ca fertilizer. It should also be noted that rock phosphate and rock phosphate-derived fertilizers, such as superphosphates, also contain Ca.

Aluminum

Exchangeable Al is frequently the dominant cation in soils

with pH less than 5.0. Such soils are usually low in Ca and Mg and may have a low effective cation exchange capacity (CEC)[1]. When the soil pH is greater than 5, the Al saturation is low. Hence the management of an acid soil in the tropics is often one of Al management, particularly if Al sensitive crops are grown. The purpose of liming is to neutralize excessive Al rather than increase soil pH.

Some soils of tropical regions have high levels of exchangeable Al. Sanchez and Isbell (1979) reported exchangeable Al values ranging from 0.1 to 7.4 me/100 g soil in Oxisols, Ultisols, and Alfisols of four South American countries with soil pH ranging from 4.0 to 5.7. Expressed in terms of Al saturation of effective CEC, the range was from 1 to 81 percent, with more than one-half of these soils showing Al saturation values of over 60 percent, a generally accepted critical level for Al toxicity for most crops (Kamprath, 1978). Many of these soils had high Al values in the subsoil, a condition that is disturbing because it is difficult to correct subsoil acidity. On soils from West Africa (mainly Ultisols and some Alfisols), Juo (1977) reported Al saturation values of 35 to 40 percent for surface soils and 50 to 60 percent for subsoils. Lopes and Cox (1977) showed that 80 percent of surface Cerrado soils in central Brazil (mainly Oxisols, Entisols, Inceptisols, and Ultisols) had more than 40 percent Al saturation.

The physiological manifestations of Al toxicity have been characterized for a number of crops, especially by Foy and his associates (1969). From data obtained from experiments on soybean and other crops, it is apparent that Al toxicity induces two changes: a decrease in the concentration of Ca in the aerial parts and accumulation of Al in the roots. Al in the soil can seriously hamper plant uptake of P and Ca. Al precipitates with phosphate and reduces its availability.

One effect of liming pasture legumes might be due to a reduction in Al availability. The response of different tropical legumes to Al concentrations varies. For example, very tolerant species include Desmodium uncinatum, Lotononis bainesii, Macroptilium lathyroides, and Trifolium rueppellianum (Andrew et al., 1973). Species of intermediate sensitivity are Trifolium subterraneum (Munns, 1965), T. repens, and Glycine wightii (Andrew et al., 1973). Al toxicity in symbiotically grown legumes has received virtually no study (Munns, 1976).

It is now generally agreed that lime recommendations for acid soils of the tropics should be primarily based on the amount of exchangeable Al. The Al saturation percentage for satisfactory plant growth varies with different crops. Kamprath (1970) reported that Al saturation at the pH of the soil in the field should be less than 45 percent for maximum growth of corn, 20 percent for soybeans, and 10 percent for cotton. He found that lime rates based on milliequivalents (me) of exchangeable Al times 1.5 resulted in neutralizing 85 to 90 percent of the exchangeable Al in

---

[1]Effective CEC is the sum of the exchangeable Ca, Mg, K, and the acidity extracted with a neutral unbuffered salt solution (Coleman, et al., 1959).

soils containing 2 to 7 percent organic matter. Studies on Cerrado soils of Brazil suggested that liming recommendations can be based on the following formulation:

T lime/ha = (2 x exchangeable Al) + [2 - me (Ca + Mg)]

This formula is believed applicable only to those soils with an effective CEC of less than 5 me/100 g (Lathwell, 1979).

Where subsoil acidity is a problem deep incorporation by plowing becomes very expensive. For immediate relief, lime should be incorporated as deeply as economically possible. The other alternative is to wait until the surface-applied lime has leached into the subsoil, with the period of waiting dependent on climatic and soil conditions.

Manganese

In some soils of the tropics, Mn toxicity limits the growth of pasture legumes. Manganese binds to NO_3-reductase and is a partial competitive inhibitor of the enzyme. Soil solution concentrations as high as 1640 ppm Mn have been reported for a Puerto Rican Oxisol of pH 4.5 (Brenes and Pearson, 1973).

The exchangeable Mn content of soil is related to soil pH and the largest decrease in exchangeable Mn occurs when acid soils are limed to pH 5.5. Mn availability is also related to oxidation-reduction conditions in the soil, e.g. Mn availability is increased when a soil is subjected to waterlogging or other reducing conditions. High temperature and excessive drying also cause an increase in Mn availability. Because of the interplay of soil reaction, oxidation-reduction, and microbiological factors, the status of exchangeable Mn is less predictable than the status of exchangeable Al. In Hawaii, for example, the most manganiferous soil is an acid (associated with high availability), well-oxidized (associated with low availability) Oxisol. On the other hand, an Utisol which is also acidic and well-oxidized, with presumably about the same microbiological and parent material backgrounds as the Oxisol, contains a low amount of available Mn. Obviously, factors, most likely environment and management related, must play a role.

Nodulation and N_2-fixation processes are more sensitive to Mn toxicity than host plant growth and Rhizobium strain differences have been observed. Critical levels of Mn for different legumes have been reported but it has been argued that the critical tolerance levels do not determine the relative tolerance of the plant to Mn toxicity. Some plants with low critical tolerance levels are less sensitive to Mn toxicity than plants having high critical tolerance levels. There may be genetic differences in plant capabilities to compartmentalize Mn in their tissues. Another factor that must be considered is the Mn/Fe ratio in plant uptake. Tolerance to high plant concentrations of manganese varies considerably among and within legume species. Andrew and Hegarty (1969) reported the following order of species tolerance: Centrosema pubescens > Stylosanthes humilis > Lotononis bainesii > Macroptilium lathyroides > Leucaena leucocephala > Desmodium uncinatum > Medicato sativa > Glycine wightii > Macroptilium atropurpureum, cv. Siratro. A similar (among common species) order was reported by Souto and Dobereiner (1969). Manganese toxic level, for 50 percent yield reduction, was 3.6μ M in Vigna unguiculata and over 42μ M for Centrosema pubescens and there was no evidence that legumes of tropical origin were more resistant than legumes of temperate origin as far as effects on the host-plant growth are concerned (Asher and Edwards, 1978). Manganese toxicity tolerance by the rhizobia seems to be large so that Mn toxicity is unlikely to inhibit their growth or survival in acid soils.

Liming can reduce Mn toxicity in legumes. The form of N applied may affect the response of plants to excess Mn. Excess Mn causes a greater reduction in yield when N is supplied as NO_3 than when it is supplied as NH_4.

Future Research Needs

1. Determine of critical levels of percent Al saturation in soils for important tropical pasture/forage legumes.
2. Develop formulations based on exchangeable Al for liming recommendations that would be applicable to a wide range of acid soils.
3. Select/breed pasture cultivars tolerant to Al and Mn toxicities.
4. Determine critical levels of Mn toxicity in soils and tropical pasture legumes.
5. Determine frequency of lime application under varying soil and climatic conditions with long-term trials.
6. Develop a reliable soil test for Mn that would correlate well with legume growth and symbiotic system performance.
7. Explore alternative materials in providing basic cations, e.g. selected crop residues, where lime is too expensive/unavailable.
8. Determine optimum nutrition balance ratios, e.g., Mn/Fe, Al/Ca, with important tropical legumes.

MOLYBDENUM AND OTHER MICRONUTRIENTS

Molybdenum.

Molybdenum (Mo) deficiency is a relatively common micronutrient problem in tropical legumes. It is also distinctive from the standpoint of relationship to soil acidity in that it is the only micronutrient that becomes more unavailable as acidity increases; the others (Zn, Cu, B, Fe, Mn) decrease in availability as alkalinity increases.

Mo deficiency is most likely to be associated with Oxisols, Ultisols, and Inceptisols (especially Andepts formed from old volcanic ash), followed by Alfisols and Entisols. The typical Mo-deficient soil is a very acid (pH 4.5-5.0) highly weathered soil with sesquioxides and kaolinite making up much of the soil mineralogy.

Mo plays a dual function in plant growth. Small quantities of Mo are needed to reduce NO_3 to NH_4 as the first step in protein synthesis, and relatively greater quantities are required in the symbiotic N_2-fixation reductase enzymes.

Liming increases Mo availability by releasing Mo from soil minerals, such as iron and aluminum oxides. Therefore, raising pH to release Mo for plant use is one way of dealing with the deficiency problem. This method may be expensive because many tropical soils are highly buffered and large amounts of liming material may be needed to raise soil pH. The second approach is to apply Mo with other fertilizers. In Australia, molybdenized superphosphate (approximately 1 part MoO_3 per 1500 parts superphosphate) is used in such a manner.

Because the amount of Mo required to correct deficiencies varies with soil, legume, method of application, and other factors, it is understandable that large differences in the rates of Mo recommended can be found in different parts of the tropics. The concentration of extractable Mo required to distinguish between deficient and non-deficient soil has been reported to range from 0.12 to 0.14 ppm by some workers, whereas others did not find a good correlation between extractable Mo and response. Mo content of legume tops can be a poor guide to an adequate supply because Mo deficiency limits growth by restricting the N_2-fixation process and Mo tends to concentrate in the legume root nodules. Anderson (1956) reviewed the subject of Mo nutrition and indicated that a good response of a legume to Mo is expected if plants contain 0.5 ppm Mo. At a concentration less than 0.1 ppm in dry matter, Mo deficiency is likely. Plants growing on low Mo soils have numerous small and ineffective nodules of white or green color, whereas with sufficient Mo, the effective nodules may be larger and pink to red inside.

The Mo requirement for optimum N-fixation and growth differs among legumes and is much smaller if the plants are supplied with combined N. In a recent report (Anon., 1978) the sensitivity of some tropical legumes to Mo deficiency was indicated as follows: Glycine > Desmodium > (Macroptilium) Siratro > Lotononis > Stylosanthes.

Both molybdic oxide (MoO_3) and sodium molybdate (Na_2MoO_4) have been extensively used by Australian workers to supply Mo. A typical application might be: 100 g Mo/ha applied every three years, with higher rates for legumes that require large amounts of Mo or on soils that strongly sorb Mo. Both carriers are not entirely satisfactory, the first because of poor solubility and the second because of a toxic salt effect on rhizobia. Although the cost of Mo fertilizer is relatively low and small quantities are required, some mixing expense is usually incurred to ensure even distribution of such small quantities.

Mo deficiency leads to an ineffective symbiotic system, whereas its excess in forage is harmful to animals. Mo deficiency can also lead to nitrate accumulation in plants which is related to nitrate-poisoning of animals. Mo excess in forage can cause copper deficiency in animals, a situation called molybdenosis or teart. Therefore, a delicate balance is needed.

Mo application cannot always be a substitute for liming. If soil pH is so low that Al or Mn toxicity or both is a problem, small amounts of lime must be applied to alleviate this Al or Mn toxicity. Because many tropical forage legumes are adapted to low pH values, the application of Mo without additional lime may be appropriate. Tang (1974) reported that in a lateritic soil of pH 4.7, an application of 60 g/ha of MoO_3 gave the same yield increase of Siratro as 2000 kg/ha of lime.

Future research activities should address the subject of "Mo in place of lime" for tropical pasture legumes to find a clear answer to the question. It is also important to characterize tropical soils for Mo adsorption and to determine critical Mo concentrations for maximum dry matter production and N_2-fixation for individual forage legumes.

Other Micronutrients (Zn, Cu, B, Mn, Fe)

The micronutrients, zinc (Zn), copper (Cu), boron (B), manganese (Mn), and iron (Fe) become less available under neutral to alkaline conditions which are often found in Vertisols, Aridisols and some Entisols, and Mollisols. Deficiency of these micronutrients caused by soil reaction is associated with these soil orders. Micronutrient deficiencies, however, are also found in acid, highly weathered soils in the tropics. Soil attributes that are often associated with deficiency are a sandy texture (natural sands or clay-aggregated "sands") and an old land-surface origin which is already low in micronutrient reserve. Furthermore, intensive agricultural practices of today with increased plant density and application of concentrated forms of fertilizer, without much return of crop residue to the soil, will contribute to micronutrient depletion.

More deficiencies have been reported in tropical countries for Zn, Cu, and B than for Fe and Mn. In spite of these reports, however, few systematic or definitive studies on the role of micronutrients in crop production are available (Drosdoff, 1972). There also appears to have been a failure to adequately characterize and classify the affected soils.

Future Research Needs

1. Systematically identify micronutrient-deficient soils and adequately characterize soils for proper classification.

2. Establish critical levels of micronutrients in soils and in legumes.

3. Carry out long-term field experiments on micronutrient corrective measures, with emphasis on maintenance requirements.

4. Include the role of crop residue return to soil as related to micronutrient recycling.

PHOSPHORUS

Phosphorus is the element most limiting in growth of legumes in tropical regions. Many highly weathered soils of the tropics, especially Oxisols, Ultisols, Andepts (a suborder of Inceptisols), and oxic Alfisols, are capable of fixing large quantities of applied P. Fixation can reduce P availability. Many believe that widespread P deficiency, and the cost of overcoming it, present an obstacle to agricultural expansion into the under-utilized and highly weathered soils of the tropics. With the great majority of tropical soils, the question is not whether to apply P but how much to apply. A follow-up question would be how to use P fertilizers most efficiently.

P Fixation in Soil

Tropical soils adsorb P in amounts which depend upon soil pH, Fe and Al oxides, organic matter, and soil clay content. Fox (1967) determined P sorption capacity of some Hawaiian soils and reported that the capacity of various mineralogical systems for fixation are as follows: amorphous hydrated oxides > gibbsite-goethite > kaolin ≯ montmorillonite. Review of the literature shows reports of high P fixing capacities in soils from almost all countries in the tropics. Two mechanisms, chemical adsorption and precipitation, are considered to be responsible for P fixation. To make P fertilizer recommendations, three steps should be taken:

1. Measure amount of P available in the soil (solution concentration).
2. Determine optimum level of available P for a particular crop.
3. Increase the amount of fertilizer from the present level to the optimum level.

This information is available by using the P adsorption isotherm approach of Fox and Kamprath (1970).
The other commonly used method of determining the P requirement of a crop is to extract soil with an acidic or alkaline extractant and relate the concentration of P in soil to crop yield. A disadvantage of using extractable P methods is that direct comparison of available P is not possible when different extraction methods are used. Therefore, the method should be specified when noting critical level of soil P.

P Requirements of Some Tropical Forage Legumes

Legumes have a high requirement for P due to the production of P containing protein. P deficiency can be observed on plant dry matter production and on N concentration in the legume. Phosphorus application can also stimulate microbial activity and increase the amount of available N. Number, density, and growth of nodules are greatly stimulated by P application.

Critical P Concentration in Plants

Critical P concentration values have been used in P fertilization programs with tropical legumes. As an example, critical concentration in the tops of nine tropical legumes at preflowering stage was found to range from 0.16 to 0.25 percent (Andrew and Robins, 1969a). However, there are limitations to the use of such values unless certain conditions are met.

Because there is a marked decline in critical percentage with an increase in age of plants, the age of the plant at sampling should be specified. It is likely that some of the variations in critical concentrations are due to differences in the plant part and its age at the time of the sampling. Therefore, it is important to standardize the sampling procedure. There is also a need to correlate P concentration in the plant with different soil types if one were to apply P fertilizer at planting in anticipation of need, especially with perennial legumes.

Type of Fertilizer and Application Method

Single superphosphate, triple superphosphate, rock phosphate, rock phosphate fused with $MgSiO$, biosuper, calcined rock phosphate, and citraphos have been used on pastures in various tropical countries.

Single superphosphate has been most commonly used because it supplies S along with P. In Brasilia, a local rock phosphate was only one-third as effective as superphosphate in the first year for Stylosanthes humilis and Brachiaria decumbens pastures whereas rock phosphate heated with $MgSiO$ and Morroccan rock phosphate had similar effectiveness as superphosphate. In the second year, the local rock phosphate was similar to superphosphate under high acid conditions. P supply from rock phosphate is limited in the first season because of low P availability. Availability increases in the second year because of higher soil acidity. Plant species differ in their effectiveness in utilizing P from rock phosphate. When Nauru rock phosphate was used, Lotononis bainesii, Centrosema pubescens, Indigofera spicata, and Stylosanthes guianensis gave higher yields with rock phosphate then did Desmodium uncinatum and Macroptilium lathyroides (Bryan and Andrew, 1971). Plants which use P efficiently generally have large root volumes and abundant root hairs. These attributes increase root interception of P. The resulting increased number of adsorption sites may explain the improved uptake of P. Other factors that may play a part in increased uptake include the ability of roots to acidify the rhyzosphere, the presence of mycorrhizae and the excretion of phosphate enzymes. In some soils that fix large amounts of P, a mycorrhizal association with the legume may be important (Mosse et al., 1976). Mycorrhizae may also increase the utilization of rock phosphate (Murdoch et al., 1967).

Biosuper is a biological P fertilizer in which the bacteria Thiobacillus oxidizes the S which acidifies the rock phosphate. It is made by pelleting 5 parts of rock phosphate, 1 part elemental S, and a small quantity of inoculum of the bacteria <u>Thiobacil-</u><u>lus</u> (Swaby, 1975). This material tends to be inferior to super-phosphate in the first year with respect to dry matter yield and P yields, but over longer periods it gives comparable results (Jones and Field, 1976).

Calcined rock phosphate is rock phosphate heated between 600 to 900°C. P from calcined rock phopshate is more available to plants than from the unheated source but the increased availabil- ity must be balanced against the increased processing cost.

Phosphorus has been broadcast applied to pastures more often than band applied. Sometimes there is an advantage to band seed- ing and fertilizer application. Band placement appeared to bene- fit the legume more than the associated grass when basic slag was used as a source of P (Anon., 1976). It has been suggested how- ever, that when P is banded, root growth may be restricted to the banded area and the plants are susceptible to drought during short periods when it does not rain. Short periods of drought are com- mon in many climates where Oxisols and Ultisols occur. There is a need for more research on P placement for pasture establishment and production in the tropics.

Future Research Needs

1. Determine of soil and plant P requirements of important tropical pasture legumes.
2. Evaluate the availability of intermediately processed forms of rock phosphate, especially in acid soils.
3. Study the long-term residual effect of P application and its influence on maintenance requirements in perennial legumes.
4. Select/breed for cultivars tolerant to low soil P.

POTASSIUM

Soils of the tropics are generally higher than their temper- ate counterparts in kaolinite, a 1:1 clay not capable of K fixa- tion, and in iron/aluminum oxides. This means a relatively K-poor soil to begin with, as well as a soil that is a poor fixer of K. Under such conditions, K deficiency can develop rapidly when land is cleared and intensively cropped, especially in high rainfall areas. Many Oxisols, Ultisols, Inceptisols, Alfisols, and Enti- sols in the tropics contain less than 0.2 meq exchangeable K, a minimal level according to some (Sanchez and Isbell, 1979). Ver- tisols, generally montmorillonitic in nature, are an exception and usually contain sufficient K.

The use of exchangeable K values as a criterion for critical level in soil, however, poses serious problems in diagnosing K- deficient areas. Non-exchangeable sources of K are known to slow- ly release available K to plants during a growing season. Fur- thermore, the analytical technique (extraction with ammonium ace- tate) used in determining exchangeable K also does not take into account the cationic environment of K in soil. In view of such problems, a more dynamic approach to express the quantity-intensity relationship involved in K availability has been used (Boyer, 1972 and Fergus et al., 1972). Recently, Singh and Jones (1975) reported the use of K sorption isotherms to determine K requirements of some plants.

As with fertilizer programs for other nutrient elements, the plant itself has been utilized to denote K sufficiency or deficiency. The critical levels of K in tropical pasture legume tops range from 0.6 to 0.9 percent (Andrew and Robins, 1969b). If K concentration is 1-2 percent, no response is expected from K fertilization. The severity of K deficiency in a pasture increases with time because the grass competes with the legume for the available K in soil. The leguminous plants which have a high cation exchange capacity (CEC) of roots, absorb divalent cations with much greater energy than monovalent cations (Fox and Kacar, 1964). The grasses have a low CEC and absorb K with greater ease. The result is that K fertilization usually increases the proportion of legume in a grass-legume mixture.

There is little information on maintenance requirements for K. Evans and Bryan (1973) recommend a 30 kg/ha/year rate and advised that it was a good practice to wait for the deficiency symptoms to appear before applying K. If KCl is used as a source of K, high rates (100-200 kg/ha) may result in seedling death because of chloride toxicity. The risk of chloride toxicity can be reduced, either by using lower levels of KCl or by replacing it with K_2SO_4 as a K source.

K fertilization needs depend on crop residue management. Some tropical plants accumulate a large concentration of K. Napier or elephant grass (Pennisetum purpureum) grown in a dominantly kaolin soil in Hawaii contained as much as 4 percent K in the dried material (Kanehiro, unpublished).

Future Research Needs

1. Maintenance requirement studies for important pasture legumes.
2. Comparative studies of K requirements of Rhizobium, the host plant and the symbiotic system.
3. Correlation of soil tests with responses under different soil and environmental conditions.
4. Possibly using K sorption isotherms to determine external requirements of tropical legumes.
5. Measure leaching loss of K in high rainfall areas under various fertilizer and management practices.

SULFUR

Sulfur deficiency may be more widespread than is presently thought. In many countries superphosphate is used as a source of P and this fertilizer contains about 10 percent S. Therefore, S is also supplied when superphosphate is applied. In recent years interest has shifted from conventionally low analysis P fertilizers to highly concentrated forms. The concentrated fertilizers may contain only low amounts of S or none at all, and S deficiency is more likely.

Typically, sulfur shortage occurs in highly weathered acid soils where intensive leaching associated with high rainfall has contributed to its removal. Sulfur deficiency, however, is not always associated with such areas because their soils have high S retention. In this way, the distribution of S deficiency differs from that of P deficiency in that with the latter, high rainfall usually means wider incidence of deficiency. The reason is that although adsorption characteristics are somewhat similar for the two nutrient elements, S is absorbed much less strongly, and thus the absorbed form of S can maintain adequate levels of S in the soil solution for plant nutrition.

There is considerable evidence that the addition of S to a S-deficient soil will increase nodulation of tropical pasture legumes and thus result in higher yield. The S requirement for maximum dry matter yield of a tropical legume was lower than that for maximum N yield (Jones and Robinson, 1970).

Two approaches are being used to characterize S status of a soil. The first is to determine extractable sulfate (0.01 M Ca_2HPO_4 is commonly used as an extractant) and relate it to plant performance. For instance, Probert and Jones (1977) indicated that 4 ppm phosphate-extractable S in soil was the critical level for several species. They cautioned that such a level varies with species and environment. The other approach is to determine critical S levels in plants, as exemplified by levels for seven tropical legumes which ranged from 0.14 to 0.17 percent, (Andrew, 1977). In some cases the latter approach gives better predictions of plant response to S than the soil-analysis approach because plant absorption is measured instead of S status. However, the plant approach has its limitations, too, because the critical value may change with age or with the plant environment.

To correct S deficiency, an application rate of 10-20 kg S/ha over a two-year period is often sufficient. Good responses to elemental S and gypsum ($CaSO_4$) have been reported throughout the tropics. Another S fertilizer is Biosuper. It has Thiobacillus added to convert elemental S to sulfate. Residual effects of S applications are generally good to excellent, provided intensive leaching conditions do not exist in the soil.

Future Research Needs

1. Collection of additional S data for a comprehensive world map on S-deficient areas.

2. Further development of soil chemical methods to estimate plant available S.

3. Refinement of index tissue selection in assessing critical levels of S in tropical legumes.

4. Collection of additional data on maintenance requirements of S.

5. Determination of S sorption-desorption characteritics of tropical soils.

6. Determination of S requirements of grass-legume mixtures.

7. Differentiation of S requirements of a legume from those of Rhizobium.

COMBINED NITROGEN

Combined nitrogen is defined as inorganic (or available) N coming from the addition of fertilizer N and from the soil itself through N mineralization.

There could be three possible results of the inorganic N supply on N_2 fixation: i.e. positive (stimulatory), negative, or no effect, and the published literature contains all three kinds of reports. If we need to make a generalization, it is safe to say that small amounts of available N in soil at the right time stimulated N_2-fixation, and large amounts are harmful for the symbiotic association. A schematic diagram of the effect of combined N on N_2-fixation is represented in Figure 2.1. Our interest is to find point "x" on the x-axis, i.e. the concentration of available N in soil to be just enough to maximize N_2-fixation. However, "x" is elusive. This amount will vary, depending upon the host-species, type of bacteria, the form and method of fertilizer application, and whether the legume is planted alone or with grass and other environmental factors.

Stimulatory Effects.

There is convincing evidence that small amounts of available N applied at the proper time can stimulate nodulation and N_2-fixation. This effect presumes that many high-yielding legumes possess a growth potential which cannot be satisfied by even the most efficient type of symbiosis (Pate, 1976). It is argued that the application of small amounts of N will satisfy the needs of the seedling before N_2-fixation has started (starter N). A healthly seedling will produce a healthy plant with an extensive root system and more sites for Rhizobium. Gates and Wilson (1974) applied 0 to 60 ppm N and 0 to 1000 kg superphosphate/ha to examin their effect on Stylosanthes humilis. High N and high P combination produced the largest and the best nodulated plants. Thus, the addition of starter N would likely to be most effective in: 1) establishing seedlings in highly weathered, low organic matter, acid soils and in 2) established but nutrient-impoverished pastures or fields coming out of heavy grazing. If possible, application should be done during the dry season before rains.

Inhibitory Effects.

Available N such as that occurring after cultivating a virgin forest soil can limit symbiotic N_2-fixation. In tropical soils, warm temperatures and moist conditions generally favor rapid nitrification if soil pH is not limiting. In some soils of Hawaii, the rate of NH_4-N transformation to NO_3-N was about 1.2 ppm/day and 80 ppm N was produced as a result of organic matter decomposition in 8 weeks (Singh and Kanehiro, 1970). High levels of NO_3-N

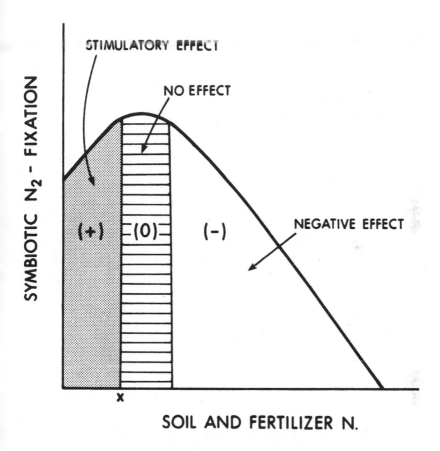

Figure 2.1.   Schematic diagram showing the effect of combined nitrogen on symbiotically fixed nitrogen.

are known to occur in some tropical soils and values ranging from 18 to 400 ppm N were reported by Norris and Date (1976).   Under these conditions, available N can accumulate to high levels even without the addition of fertilizer N.

Nitrates inhibit the production and curling of root hairs and may inhibit the initiation and development of infection threads. There are two possible explanations for NO_3-N inhibition of infection by rhizobia.   1) Nitrates reduce the synthesis of indoleacetic-acid (IAA) which is necessary for nodule initiation.   Rhizobium produces NO_2 from NO_3 which possibly oxidizes IAA.   2) In the presence of NO_3, photosynthate is used in the shoots and roots for the assimilation of nitrate and its supply to nodules is decreased, affecting the symbiotic system adversely.

Munns (1968) showed that 0.02 to 0.05 mM N concentration in

solution inhibits nodulation at the beginning but not in the later stages of plant growth. NH_4-N seems to affect nodulation less than does NO_3-N. Urea and organic matter in turn are less inhibitory than NH_4- and NO_3-N. Nitrate is toxic to the Rhizobium.

Deep placement of N fertilizer has also been suggested to overcome the inhibitory effects of combined N on nodulation. To avoid undue stimulation of weeds and grasses, pelleting the legume seed with a low level of combined N has also been suggested.

Of late, more information on inoculation requirements of tropical pasture forage legumes is being published, e.g. Date (1976) and Halliday (1979).

Future Research Needs

1. Determination of the minimum amount of N needed in soil to maximize N_2-fixation for tropical pasture legumes.

2. Determination of the effects of inorganic N on Rhizobium.

3. Determination of the aspect of symbiosis most sensitive to NO_3 levels in tropical forage legumes.

4. Establishment of a compromise between fertilizing for maximum N_2-fixation and total N yield.

SOIL SALINITY

Salinity problems occur when excessive concentrations of soluble salts accumulate in the soil. These are the chloride, sulfate, and bicarbonate salts of calcium, magnesium, and sodium. Fortunately, in the tropics, large areas are made up of well-aggregated soils, such as Oxisols, Ultisols, and Inceptisols. In such soils permeability is great and salts move easily, removing salts from the plow layer and minimizes salinity problems. There are a number of soils however, that present salinity problems, such as Vertisols. Some Australian and Indian Vertisols are characterized by moderately high salt levels. In South America, large areas of mostly salty swamp, are found along the northern coast and the southern part of the east coast, especially at the mouth of the Amazon. Similarly, swamplands located near the mouths of large rivers in Africa, e.g. Casamance River in Senegal, pose salinity problems. In South and Southeast Asia coastal saline soils, not necessarily associated with rivers, add up to millions of hectares. Salinity is also a major problem in irrigated desert areas of the tropics. Low rainfall areas with low relief are prone to salinization.

Tropical legumes are generally less tolerant of salinity than temperate legumes. An investigation of salt tolerance of most of the important tropical pasture legumes and Medicago sativa revealed the following order of tolerance: Medicago sativa > Macroptilium atropurpureum > Macroptilium lathyroides > Desmodium intortum > Macrotyloma uniflorum > Vigna unguiculata > Glycine wightii > Lotononis bainesii > Lablab purpureus > Stylosanthes humilis > Desmodium uncinatum. Desmodium uncinatum and Desmodium intortum are sensitive to chloride excess. A chloride concentration of 2

percent in Desmodium intortum was associated with seedling death
(Andrew and Robins, 1969b). Little information is available on
the effect of salinity on rhizobial survival and effectiveness.

Future Research Needs

1. Screening and selection of legume cultivars for salt
tolerance.
2. Determination of critical levels for salinity tolerance
at different ages for important tropical pasture legumes.
3. Investigation of the effects of salinity on survival of
different rhizobia and the symbiotic system.

SOIL TEMPERATURE

High soil surface temperatures can occur in the tropics; a
soil surface temperature as high as 50°C has been recorded in
Zaire (Masefield, 1958). Norris (1970) pointed out that in many
areas, soil temperature at 2.5-5.0 cm depth can range from 40 to
45°C for up to 6 hours a day. At a depth of 150 cm mean annual
temperature remains constant at about 25 to 26.5°C. Very limited
data are available on actual soil temperature measurements
throughout the world. Chang (1958) prepared soil temperature maps
of the world and gave an average annual range of values for the
tropics. Average values, however, may not be good enough to show
the effect of temperature on legumes and their symbiotic system as
large temperature variations for even a short period may kill the
Rhizobium or the legume or both.
Whereas soil temperatures are measured only at a few locations and records are frequently of brief duration, air temperature observations are made daily at many sites throughout the
world. Attempts have therefore been made to predict soil temperatures from air temperatures. Toy et al. (1978) used simple linear
models for estimating mean annual, seasonal, and monthly soil temperatures with reasonable accuracy, using only air temperature
data collected by the National Weather Service at stations
throughout the United States. Better accuracy in prediction might
be obtained with additional climate and soil variables. For example, cloud data would be of great value because it affects solar
radiation which in turn affects soil temperature.
Four aspects of temperature are of interest to the legume-
Rhizobium symbiosis and legume growth and will be discussed.

1. Temperature effects on legume growth and composition.
2. Temperature effects on nitrogen fixation.
3. Temperature effects on nodulation.
4. Survival of rhizobia in soil.

Effect of Temperature on Legume Growth and Composition

Tropical pasture legumes seem to fall into at least two
groups, referred by Sweeney and Hopkinson (1975) as the "warm

tropical legumes" and the "cool tropical legumes." The first group includes Macroptilium, Centrosema, Pueraria, and Stylosanthes and appears to have a yield plateau above 27°C air temperature. The "cool tropical legumes" group is characterized by better production at lower temperatures and significant decrease in growth above 27°C air temperature. This group includes Desmodium intortum, Desmodium uncinatum, Glycine wightii, Desmodium sandwicense, Macrotyloma axillare, and Lotononis bainesii. Lotononis bainesii is known to tolerate frost. At low temperatures, seed formation in Stylosanthes humilis may be adversely affected. In New Guinea, sowing of Lablab purpureus (Dolichos lablab) for seed production is not recommended where minimum temperatures exceed 18°C. For Centrosema pubescens the minimum temperature should exceed 13°C for appreciable growth. Varietal differences in temperature tolerances have been observed in many tropical pasture legumes. Sowing dates can be manipulated to increase dry matter yields.

Desmodium intortum is most tolerant of shade and Macroptilium atropurpureum cv. Siratro the least, whereas Centrosema pubescens and Glycine wightii are intermediate.

High temperatures generally decrease the dry matter digestibility of growing herbage by increasing rate of plant development and consequent synthesis of stem material. The concentrations of mineral nutrients, especially of Mg, Ca, Zn, and B, in herbage often increase at higher soil temperatures.

Probably the best solution to the detrimental influence of temperature extremes on N fixation in legumes is to find legumes suitable for a specific environmental condition.

Effect of Temperature on Nitrogen Fixation

Nitrogen fixing activity of a nodulated legume plant exhibit a very wide temperature tolerance and shows an optimum in activity, spanning a 15-20°C temperature range. The lower limit for N-fixation in tropical legumes seems to be more affected by the environment than the plant and N_2-fixation appears to be less sensitive to temperature than nodulation.

Effect of Temperature on Nodulation

Infection, nodule initiation, and nodule development stages can be influenced by temperature of the environment. among these, root hair infection and development of bacteroid tissue appear to be most temperature sensitive. There are few studies on the effects of root temperature on nodulation and N_2-fixation in tropical pasture legumes.

For nodulation of Glycine wightii, Desmodium intortum, Desmodium uncinatum, Stylosanthes humilis, and Macroptilium atropurpureum, 30°C root temperature seems to be the best. Older plants are more tolerant of lower shoot temperatures than younger plants, presumably due to a greater effect of lower shoot temperature on nodule formation than on nodule function.

Survival of Rhizobia

The rhizobia must survive in soil, in culture media, and on seed under variable climatic conditions prior to infecting the legume. Generally high temperatures are of more concern in the survival of rhizobia in the tropics than low temperatures. In several studies, temperature was found to be a limiting factor in rhizobia survival in soil. Survival is affected by soil type, exposure time, Rhizobium strain, moisture status, and initial concentration of inoculum. The maximum temperature range for the growth of rhizobia from the cowpea miscellany group seems to be 30-42°C. The exposure of rhizobia to 40°C for 3-8 hours may result in a ten-fold reduction in their number (Bowen and Kennedy, 1959). Some types of 2:1 clay provide protection of rhizobia against high temperatures, probably due to a protective mechanism of a clay envelope around the bacteria which decreases the loss of water from the bacterial cells. To improve rhizobia survival under high temperature conditions, the following have been suggested: a) develop rhizobia resistant to high temperatures, b) plant inoculated seed at the proper depth, c) sow inoculated seeds in later part of the day, d) use larger amounts of inoculum, and e) use mulch to modify environment.

Lowendorf (1977) recently studied rhizobia survival under different conditions and concluded that when temperature was high survival of rhizobia was greater in dry soil than in moist soil.

Future Research Needs

1. Developing a comprehensive soil temperature map of the world.
2. Screening of rhizobia for different climatic extremes.
3. Breeding of legumes to suit different environmental conditions.

REFERENCES

Anderson, A.J. 1956. Molybdenum as a fertilizer. Advances in Agronomy 8: 163-202. American Society of Agronomy, Madison, Wisconsin.

Andrew, C.S. 1976. Effect of calcium, pH, and nitrogen on the growth and chemical composition of some tropical and temperate pasture legumes. I. Nodulation and growth. Australian Journal of Agricultural Research 27: 611-623.

Andrew, C.S. 1977. The effect of sulfur on the growth, sulfur and nitrogen concentrations and critical sulfur concentrations of some tropical and temperate pasture legumes. Australian Journal of Agricultural Research 28: 807-820.

Andrew, C.S. 1978. Mineral characterization of tropical forage legumes. In "Mineral Nutrition of Legumes in Tropical and Subtropical Soils," (eds.) C. S. Andrew and E. J. Kamprath, CSIRO, Melbourne, pp. 93-111.

Andrew, C.S. and Hegarty, M.P. 1969. Comparative responses to manganese excess of eight tropical and five temperate pasture legume species. Australian Journal of Agricultural Research 20: 687-696.

Andrew, C.S., Johnson, A.D., and Sandland, R.L. 1973. Effect of aluminum on the growth and chemical composition of some tropical and temperate pasture legumes. Australian Journal of Agricultural Research 24: 325-339.

Andrew, C.S. and Norris, D.O. 1961. Comparative responses to calcium of five tropical and four temperate pasture legume species. Australian Journal of Agricultural Research 12: 40-55.

Andrew, C.S. and Robins, M.F. 1969a. The effect of phosphorus on the growth and chemical composition of some tropical pasture legumes. I. Growth and critical percentages of phosphorus. Australian Journal of Agricultural Research 20: 665-674.

Andrew, C.S. and Robins, M.F. 1969b. The effect of potassium on the growth and chemical composition of some tropical and temperate pasture legumes. I. Growth and critical percentages of potassium. Australian Journal of Agricultural Research 20: 999-1007.

Anon. 1976. Pastures and forages. Beef Production Program, CIAT Annual Report 1976 Cali.

Anon. 1978. Molybdenum needs of tropical legumes. Rural Research 98. CSIRO, Melbourne.

Asher, C.J. and Edwards, D.G. 1978. Relevance of dilute solution culture studies to problems of low fertility tropical soils. In "Mineral Nutrition of Legumes in Tropical and Subtropical Soils," (eds.) C. S. Andrew and E. J. Kamprath, CSIRO, Melbourne.

Boyer, J. 1972. Soil potassium. In "Soils of the Humid Tropics." National Academy of Sciences, Washington, D.C. pp. 102-135.

Bowen, G.D. and Kennedy, M.M. 1959. Effect of high soil temperature on Rhizobium spp. Queensland Journal of Agricultural Science 16: 177-197.

Brenes, E. and Pearson, R.W. 1973. Root responses of three Gramineae species to soil acidity in an Oxisol and an Ultisol. Soil Science 116: 295-302.

Bryan, W.W. and Andrew, C.S. 1971. The value of Nauru rock phosphate as a source of phosphorus for some tropical pasture legumes. Australian Journal of Experimental Agriculture and Animal Husbandry 11: 532-535.

Chang, J.H. 1958. Ground temperature. Vol. 1. Harvard Univ. Blue Hill Meterological Observatory, Milton 86, Massachusetts.

Coleman, N.T., Weed, S.B., and McCracken, R.J. 1959. Cation-exchange capacity and exchangeable cations in Piedmont soils of North Carolina. Soil Science Society of America, Proceedings 23: 146-149.

Date, R.A. 1976. The development and use of legume inoculants. In "Biological Nitrogen Fixation in Farming Systems of the Tropics," (eds.) A. Ayanaba and P. J. Dart, John Wiley & Sons, Chichester.

Dobereiner, J. and Aronovich, S. 1965. Effecto da calagem e da temperatura do solo na fixacao de nitrogenio de Centrosema pubescens Benth. em solo com toxidez de manganes. Proceedings IX International Grasslands Congress 2: 1121-1124.

Dradu, E.A.A. 1974. Soil fertility studies on loam soils for pasture development in Uganda: pot experiments. East African Agriculture and Forestry Journal 40: 125-131.

Drosdoff, M. 1972. Soil micronutrients. In "Soils of the Humid Tropics." National Academy of Sciences. Washington, D.C.

Evans, T.R. and Bryan, W.W. 1973. Effects of soils, fertilizers and stocking rates on pastures and beef production on the wallum of southeastern Queensland. 2. Liveweight change and beef production. Australian Journal of Experimental Agriculture and Animal Husbandry 13: 530-536.

Fergus, E.F., Martin, A.E., Little, I.P. and Haydock, K.P. 1972. Studies on soil potassium. II. The Q/I relation and other parameters compared with plant uptake of potassium. Australian Journal of Soils Research 10: 95-111.

Fox, R.L. and Kacar, B. 1964. Phosphorus mobilization in a calcareous soil in relation to surface properties of roots and cation uptake. Plant and Soil 20: 319-330.

Fox, R.L. 1967. Phosphorus fixation by Hawaiian soils and what to do about it. Proceedings First Annual Hawaii Fertilizer Conference. Honolulu. pp. 28-41.

Fox, R.L. and Kamprath, E.J. 1970. Phosphate sorption isotherms for evaluating the phosphate requirement of soils. Soil Science Society of America, Proceedings 34: 902-907.

Foy, C.D., Fleming, A.L., and Armiger, W.H. 1969. Aluminum tolerance of soybean varieties in relation to calcium nutrition. Agronomy Journal 61: 505-511.

Gates, C.T. and Wilson, J.R. 1974. The interaction of nitrogen status and nodulation of Stylosanthes humilis. H.B.K. (Townsville stylo) Plant and Soil 41: 325-333.

Halliday, J. 1979. Field responses by tropical forage legumes to inoculation with Rhizobium. In "Pasture Production in Acid Soils of the Tropics," (eds.) P.A. Sanchez and L. E. Tergas. CIAT. Cali.

Jones, R.K. and Field, J.B.F. 1976. A comparison of biosuper and superphosphate on a sandy soil in the monsoonal tropics of north Queensland. Australian Journal of Experimental Agriculture and Animal Husbandry 16: 99-102.

Jones, R.K. and Robinson, P.J. 1970. The sulfur nutrition of Townsville lucerne (Stylosanthes humilis). Proceedings XI International Grassland Congress, Surfers' Paradise 1970: 377-380.

Juo, A.S.R. 1977. Soluble and exchangeable aluminum in Ultisols and Alfisols in West Africa. Communications in Soil Science and Plant Analysis 8: 17-35.

Kamprath, E.J. 1970. Exchangeable aluminum as a criterion for liming leached mineral soils. Soil Science Society of America, Proceedings 34: 252-254.

Lathwell, D.J. 1979. Crop response to liming of Ultisols and Oxisols. Cornell International Agricultural Bulletin 35, Ithaca, New York.

Loneragan, J.F. 1973. Use of isotopes for study of fertilizer utilization in crops. International Atomic Energy Agency, Vienna.

Lopes, A.S. and Cox, F.R. 1977. A survey of the fertility status of soils under "cerrado" vegetation in Brazil. Soil Science Society of America, Proceedings 41: 742-747.

Lowendorf, H.S. 1977. State-of-the-art (SOTA) on "Survival of Rhizobium in Soil". 58 pp. (mimeo) Cornell University, Ithaca, New York.

Masefield, G.B. 1958. Some factors affecting nodulation in the tropics. In "Nutrition of the Legumes." (ed.) E.G. Hallsworth. Butterworths Scientific Public. London. pp. 202-212.

Mears, P.T. and Barkus, B. 1970. Response of Glycine wightii to molybdenized superphosphate on a krasnozem. Australian Journal of Experimental Agriculture and Animal Husbandry 10: 415-425.

Mosse, B., Powell, C.L., and Hayman, D.S. 1976. Plant growth responses to vesicular-arbuscular mycorrhiza. IX. Interactions between V. A. mycorrhiza, rockphosphate and symbiotic nitrogen fixation. New Phytologist 76: 331-342.

Munns, D.N. 1965. Soil acidity and growth of a legume. II. Reactions of aluminum and phosphate in solution and effects of aluminum, phosphate, calcium, and pH on Medicago sativa L. and Trifolium subterraneum L. in solution culture. Australian Journal of Agricultural Research 16: 743-755.

Munns, D.N. 1968. Nodulation of Medicago sativa in solution culture. III. Effects of nitrate on root hairs and infection. Plant and Soil 29: 33-47.

Munns, D.N. 1976. Soil acidity and related factors. In "Exploiting the Legume-Rhizobium Symbiosis in Tropical Agriculture." Proc. of a Workshop at Kahului, Maui, (eds.) J.M. Vincent, A.S. Whitney, and J. Bose. University of Hawaii, College of Tropical Agriculture. Miscellaneous Pubication 145 pp. 211-236.

Munns, D.N., Fox, R.L., and Koch, B.L. 1977. Influence of lime on nitrogen fixation by tropical and temperate legumes. Plant and Soil 46:591-601.

Norris, D.O. 1970. Nodulation of pasture legumes. In "Australian Grasslands," (ed.) R. M. Moore. Australian National University Press, Canberra.

Norris, D.O. and Date, R.A. 1976. Legume bacteriology. In "Tropical Pasture Research, Principles and Methods," (eds.) N. H. Shaw and W. W. Bryan. Commonwealth Bureau of Pastures and Field Crops Bulletin 51. pp. 134-174.

Pate, J.S. 1976. Physiology of the reaction of nodulated legumes to environment. In "Symbiotic Nitrogen Fixation in Plants," (ed.) P. S. Nutman. Cambridge University Press.

Pate, J.S. 1977. Functional biology of dinitrogen fixation by legumes. In "A Treatise on Dinitrogen Fixation," (eds.) R. W. F. Hardy and W. S. Silver, John Wiley and Sons, New York.

Pearson, R.W. 1975. Soil acidity and liming in the humid tropics. Cornell International Agriculture Bulletin 30. Cornell University, Ithaca, New York.

Probert, M.E.P. and Jones, R.K. 1977.   The use of soil analysis for predicting the response to sulfur of pasture legumes in the Australian tropics.   Australian Journal of Soil Research 15: 137-146.

Sanchez, P.A. and Isbell, R.F. 1979.   A comparison of the soils of tropical Latin America and tropical Australia.   In "Pasture Production in Acid Soils of the Tropics," (eds.) P.A. Sanchez and L.E. Tergas, CIAT, Cali. pp. 25-54.

Singh, B.R. and Jones, J.P. 1975.   Use of sorption isotherms for evaluating potassium requirements of plants.   Soil Science Society of America, Proceedings 39: 881-886.

Singh, B.B. and Kanehiro, Y. 1970.   Changes in available nitrogen content of soils during storage.   Journal of the Science of Food and Agriculture 21: 489-491.

Souto, S.M. and Dobereiner, J. 1969.   Manganese toxicity in tropical forage legumes.   Pesquisas Agropecuaria Brasiliera 4: 129-138.

Swaby, R.J. 1975.   Biosuper - biological superphosphate.   In "Sulphur in Australian Agriculture."   Sydney University Press. Sydney. p. 213.

Sweeny, F.C. and Hopkinson, J.M. 1975.   Vegetative growth of nineteen tropical and sub-tropical pasture grasses and legumes in relation to temperature.   Tropical Grasslands 9: 209-217.

Tang, C.N. 1974.   Effect of lime and molybdenum on the yield of siratro on lateritic soil.   Taiwan Agriculture Quarterly 10: 112-119.

Toy, T.J., Kuhaida, Jr., A.J., and Munson, B.E. 1978.   The prediction of mean monthly soil temperature.   Soil Science 126: 181-189.

Trigoso, R. and Fassbender, H.W. 1973.   Effects of the applications of calcium plus magnesium, phosphorus, molybdenum, and boron on the production and nitrogen fixation in four tropical legumes.   Turrialba 23: 172-180.

3
Microbiological Considerations— *Rhizobium* Specificity for Nodulation and Nitrogen Fixation

R. A. Date

To be successful, forage legumes must be able to meet their own nitrogen requirements by reducing atmospheric nitrogen in the nodules formed by the appropriate root nodule bacteria, Rhizobium. The level of the nitrogen contribution by the Rhizobium, usually termed the "effectiveness" of the association, is an expression of the genetic compatibility between the host plant and the bacterium. The efficiency of this symbiosis is affected by environmental factors. In practice, the ability of the association to contribute to the plant's nitrogen supply varies tremendously: it ranges from a parasitic situation (i.e. the dry weight yields of the inoculated plants are less than those of the uninoculated controls) to one in which plants depending upon symbiotically fixed nitrogen give higher yields than plants growing on more than adequate levels of fertilizer nitrogen.

Rhizobia that are able to form nodules and fix nitrogen in association with the genera Centrosema, Desmodium, and Stylosanthes are part of what the literature refers to as the "cowpea-type" group. This association tends to suggest that the three genera, and the species they contain, are all nodulated effectively by the same wide range of rhizobia whereas this is partially true it is by no means entirely so; the genera display significant specificity for the level of effectiveness. For example, the wide-spectrum strain CB 756 forms effective associations on many species and species forms in all three of the genera under review. The level of effectiveness, however, varies greatly and there are significant differences both within and among species. Similarly, with a given accession, the level of effectiveness is dependent upon the rhizobial strain used. This can best be demonstrated by reference to the results obtained from a program designed to screen new accessions of Stylosanthes for their effectiveness with a wide range of rhizobia (Date and Norris, 1979).

In S. guianensis, 43 of the 67 accessions used nodulated effectively with a wide range of Rhizobium; the remaining 24 were effective with a much more limited number of strains. The number of strains effective on any one accession ranged from 0 to 17 (of the 22 strains used). There were 9 accessions ineffective with all strains and 29 effective with more than 10 strains. Similarly, for S. hamata 11 of the 37 accessions were generally nodulated effectively by a wide range and 26 by a limited range of strains. The actual number ranged from 0 to 18 (5 with no effective strains and 10 with 10 or more). These data illustrate the wide diversity of the genetic compatibility between accessions of Stylosanthes and strains of Rhizobium.

By applying pattern seeking methods it was possible to simplify the overall picture. Accessions were classified according to their ability to form effective associations with the different rhizobia. It was found that members of the groups defined possessed certain soil and climate characteristics in common. For instance, those accessions of S. hamata which had been found on alkaline soils nodulated effectively only with strains of rhizobia from alkaline soils (Date, Burt, and Williams, 1979). In S. guianensis two distinct accession groups were apparent. One of these nodulated effectively with only a limited number of rhizobial strains; it contained accessions from southern Brazil (Minas Gerais and Sao Paulo), where they were found on soils of low pH and high aluminium levels. The other group, effective with a much wider range of strains, came from other Latin American areas of mostly higher rainfalls and longer growing seasons.

It is known that the distribution of various Stylosanthes species and species types is related to both climatic and edaphic conditions (see Section III-4). The work just described suggests that rhizobial strains may be under similar constraints, possibly in association with the plants concerned. To explore this possibility rhizobia were isolated from Stylosanthes species found on contrasting soils which varied from the infertile, acid (pH 4.5) soils with high aluminium saturation found on the Colombian Llanos to the heavy clay, cracking alkaline soils of Antigua or Western Ecuador. Rhizobia from the former had a pH optimum for growth of near 4.5, considerably less than the commonly accepted optimum range of 6.5 to 7.0 (Date and Halliday, 1979). Although confirmatory work is as yet lacking it seems that the rhizobia, like their host plants, are well adapted to the edaphic situations in which they are found.

Effectiveness responses of Stylosanthes species other than those already named are poorly understood. In Desmodium relevant information is even more limited. The data of Diatloff (1968) suggested, however, that species can again be grouped according to their effectiveness with a given strain of Rhizobium; D. discolor, D. gyroides (now Codariocalyx gyroides), D. intortum, D. polycarpum, D. rensonii, D. rhytidophyllum, D. tortuosum, D. triflorum, and D. uncinatum were nodulated effectively by strain CB627 (isolated from D. intortum) but it was ineffective on D. cuneatum, D. distortum, D. heterophyllum, and D. sandwicense. More recent data (Anon. 1978) suggests significant incompatibility between D. heterophyllum and D. heterocarpum and between both of these and the D. intortum type group. D. distortum gave a mixed response being either effectively or ineffectively nodulated by a range of strains from D. barbatum, D. distortum, D. intortum, and D. uncinatum.

Similarly, for Centrosema the range of information is limited. Most examples of host-genotype variations for effectiveness of nitrogen fixation have been for C. pubescens e.g. Bowen (1959), Bowen and Kennedy (1961), and Franco, Serpa, and Souto (1973). Most strains for C. pubescens are also effective with C. plumieri and ineffective with C. virginianum. A C. virginianum x brasilianum hybrid was also ineffectively nodulated by C. pubescens strains, with the exception of the black nodule strain C101a (also referred to as CB1923 or CIAT 49) from Brazil (Anon., 1978). Strain C101a was ineffective on four accessions of C. brasilianum (H.V.A. Bushby, pers. comm.). The hybrid appears to have inherited the C. virginianum nodulation/effectiveness characteristic(s). C. brasilianum and C. virginianum have distinct and separate requirements to that for the C. pubescens types.

There is much more that we need to know about Rhizobium requirements and interrelationships for the genera Centrosema, Desmodium, and Stylosanthes. The large amount of within-genera and within-species variation for strains of Rhizobium forming effective nitrogen fixing associations precludes any attempt to group accessions purely on the basis of species or genera. An alternative would be to disregard genera and species and consider the effectiveness response pattern of individual accessions against a wide range of strains of Rhizobium. Such a system (Date, 1977) recognizes three fundamental response types:

GROUP PE--promiscuous and effective: plants nodulate effectively with a wide range of strains from different genera and species within the group; symbiotic effectiveness (host dry weight) is usually > 80 percent that of the nitrogen control, and only a few strains are not fully effective. This group corresponds most closely to that usually referred to as the 'cowpea group' and includes the genera Arachis, Calopogonium, Cajanus, Canavalia, Clitoria, Crotalaria, Cyamopsis, Desmanthus, Dolichos, Galactia, Glycine, Indigofera, Lablab, Macroptilium, Macrotyloma, Pueraria, Rhynchosia, Mucuna (Stizolobium), Teramnus, Tephrosia, Vigna, and some species of Stylosanthes and Zornia.

GROUP PI--promiscuous but often ineffective: plants frequently nodulate with a wide range of strains of Rhizobium, but many are ineffective in nitrogen fixation. Genera in this group can be further subdivided: Rhizobium for one subgroup does not always form nodules in another and those strains that do form nodules have a limited range of effectiveness. Further subgrouping based on species within genera is frequently observed, especially in Desmodium and Stylosanthes. Other genera in this group include Adesmia, Aeschynomene, Centrosema, Psoralea, and Sesbania.

GROUP S--specific: those genera and species nodulating effectively with a narrow or restricted range of strains of Rhizobium which generally originate only from nodules of homologous species or accessions. Both genera, and species within genera, usually form distinct subgroups. A large proportion of members of this group do not nodulate with strains from other genera or species and many plants nodulate irregularly when the same strain is used on different occasions. The genera most commonly cited in this group are Leucaena, Lotononis, Mimosa, Trifolium, and some species of Stylosanthes (e.g. S. capitata).

The distribution of relative effectiveness for 25 strains and a typical member of each of the above groups is illustrated in Figure 3.1. Plant dry weight and total nitrogen content are positively correlated in clovers, medics, soybeans, and lupins (Erdman and Means, 1952), and in several tropical forage legume species (D.O. Norris and R.A. Date unpublished). As specificity for effectiveness increases (i.e. from Group PE to Group S), a greater proportion of strains of Rhizobium form parasitic associations, or do not (-) or only irregularly (+) nodulate other hosts in the group (Figure 3.1).

In this way the variation in response types with a genus or species can be accommodated. For example, Stylosanthes erecta and some S. guianensis, and some S. hamata would fall into Group PE, whereas the fine stem stylo type of S. guianensis and S. sp. aff. hamata are located in Group PI. S. capitata, S. calcicola, some S. guianensis, and some S. hamata are best located in Group S (Date and Norris, 1979).

Species in Group PE will be effectively nodulated in most instances by "native" strains of Rhizobium although responses under stress conditions are likely; e.g. Stylosanthes guianensis CIAT 136 responded to inoculation and to lime pelleting when sown in the infertile acid soils of the Colombian Llanos (Anon, 1978). The species of Group S will always respond to inoculation if sown in new areas. Group PI is the most troublesome and it is appropriate to determine the need for inoculation for new sowings by field experimentation as suggested by Date (1977).

It is obvious that a knowledge of Rhizobium requirements for effective nodulation is restricted to those species of agro-economic value. There have been very few detailed appraisals of the responses of individual genera. As both traditional and less exploited legumes are being studied in greater detail and as the range of soil and climatic conditions expands (and frequently become more severe), more and more favorable responses to inoculation with effective strains are being observed. It is imperative under such conditions that the likely effectiveness response of species of interest be well documented and suitable effective strains be made available.

REFERENCES

Anon. 1978. Annual Report for 1977. Beef Program, Centro Internacional de Agricultura Tropical (CIAT) Colombia.
Bowen, G.D. 1959. Specificity and nitrogen fixation in the Rhizobium symbioses of Centrosema pubescens. Queensland Journal of Agricultural Science 16: 267-281.
Bowen, G.D. and Kennedy, M.M. 1961. Heritable variation in nodulation of Centrosema pubescens. Queensland Journal of Agricultural Science 181: 161-170.

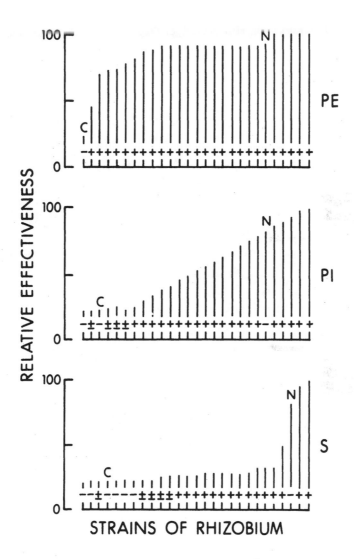

Figure 3.1. Patterns of response to an array of 25 strains of
Rhizobium for effectiveness in nitrogen fixation for
a typical member of each of the generalized effectiveness Groups PE (promiscuous and effective), PI
(promiscuous and ineffective) and S (specific).
(Relative effectiveness = dry weight of whole plant
expressed as percentage of plus nitrogen control (N).
C = uninoculated control, - = not nodulated, + =
sometimes nodulated, + = nodulated.) From Date and
Halliday (1980).

Date, R.A. 1977. Inoculation of tropical pasture legumes. In "Exploiting the legume - Rhizobium symbiosis in tropical agriculture." (eds.) J.M. Vincent, A.S. Whitney and J. Bose University of Hawaii, College of Tropical Agriculture Misc. Publ. 145 pp. 293-311.

Date, R.A., Burt, R.L., and Williams, W.T. 1979. Affinities between various Stylosanthes as shown by rhizobial, edaphic and geographic relationships. Agro-ecosystems 5: 57-67.

Date, R.A. and Halliday, J. 1979. Selecting Rhizobium for acid, infertile soils of the tropics. Nature 277: 62-64.

Date, R.A. and Norris, D.O. 1979. Rhizobium screening of Stylosanthes species for effectiveness in nitrogen fixation. Australian Journal of Agricultural Research 30: 85-104

Date, R.A. and Halliday, J. 1980. Relationships between Rhizobium for tropical legumes. In "Advances in Legume Science." (eds.) R. Summerfield and H. Bunting. Proceedings, International Legume Conference, Kew, U.K. July-August 1978. pp. 597-601.

Diatloff, A. 1968. Nodulation and nitrogen fixation in some Desmodium species. Queensland Journal of Agriculture and Animal Science 25: 165-167.

Erdman, L.W. and Means, U.M. 1952. Use of total yield for predicting nitrogen content of inoculated legumes grown in sand culture. Soil Science 73: 231-235.

Franco, A.A., Serpa, A., and Souto, S.M. 1973. Simbiose de estirpes homologas com linhagens de Centrosema pubescens. Pesquisa Agropecuaria Brasiliera, Serie Zootecnia 8: 13-17.

4
Plant-Animal Relationships

R. M. Murray

In previous sections note has been made of some of the factors which limit the enormous potential of the tropics for animal
production; low quality forage, instabilty of pasture systems, and
shortage of grazing lands have all been mentioned. Understandably, emphasis has been placed on plant factors, and the role of
plants, particularly legumes, in alleviating these problems has
been well documented. Little has been said about the animal component of the system although such a component is clearly of importance. It would, however, be inappropriate to attempt to cover
this subject in any depth and, indeed, such coverage has been presented elsewhere. Here the primary concern is to indicate research areas where it is believed a better understanding of plant-
animal interrelationships is clearly necessary. To do this, tropical grazing lands are somewhat arbitrarily separated into those
found in agriculturally intensive and extensive areas.
In intensive areas the traditional approach to the problem of
low forage quality or quantity or both is to establish better
quality grasses, to apply fertilizer and, often, to irrigate.
Pangola grass (Digitaria decumbens) is well known in the tropics
for its ability to produce high liveweight gains. Although appropriate in some areas this is an expensive process, and alternatives have been sought. Legume-based pastures is one of these.
There are, also, other approaches which appear promising. One is
to use pasture legumes in the cropping system: they help to improve soil fertility and increase the quality of the post-harvest
grazing (the quality of such grazing could, of course, also be
improved by sowing cereals with more nutritious straw). Another
approach is to use the crop itself, possible in times of over-
production, for example sugar cane provides an energy-rich diet
(Preston and Leng, 1978) to which protein rich legume supplements
can be added. In the Caribbean, Keoghan (pers. comm.) is developing this idea to suit local conditions. Here climbing legumes are
sown with sugar cane and grasses, and fertilized with pig and
poultry manure; the resultant feed is then fed back to the animals
in the dry season, thus reducing the demand for expensive supplements. Yet another approach is to sow legumes along the roadsides, areas which traditionally provide feed for draught animals.

Such systems may, or may not be, feasible in more extensive areas, although even in these areas it is common practice to fence and fertilize an area to produce a special purpose pasture for the preferential feeding of selected classes of animals. The three practices most commonly used to sustain animal production are the establishment of improved pastures (with or without fertilizer), the feeding of supplements, and the movement of animals to areas where food supply is less limiting. The primary concern is with the two former options. One may simply note that the latter, although sometimes deemed to be unacceptable, can be a very efficient way to utilize heterogenous tropical environments.

The generally accepted philosophy behind the use of a legume-based pasture is that once mineral deficiencies have been corrected, the legume fixes nitrogen which ultimately benefits the growth of the associated grass. The association then provides a better diet for the animals. It has become increasingly apparent, however, that at least some tropical legumes possess characteristics which might cause one to question this "standard" approach. In this connection, one may present three examples. First, some are capable of producing extremely high seed yields which constitute a nutritious dry season supplement (see, for instance, Playne, 1969). On the acid, infertile soils of the Llanos, Grof (pers. comm.) has recorded seed yields from Stylosanthes capitata which equal those from the average crop of soybeans. They are, however, produced at the expense of vegetative growth. Little or nothing is known about the annual feeding value of such seed crops, the extent to which they are utilized, the dependence upon the seed retention or otherwise by the plant, and the degree to which the provision of such a supplement would allow the animal to utilize the associated low quality grass herbage. A second example concerns Stylosanthes scabra, some forms of which can thrive on soils too infertile for other species (see Section II-4). Such plants are understandably low in P (phosphorus), although they have a relatively high protein content, and the effects of this unbalanced diet on animal production are unknown. Almost certainly P levels will be too low and supplementation may be necessary. This is often logistically difficult and other, more practical solutions could be sought. One approach might be to allow animals periodic access to fertilized, special purpose pastures on which the phosphorus store within the animals would be built up (in the blood and bone); this store could then be drawn upon while grazing the more extensive systems. Little is known about such aspects of animal nutrition. Finally, it is noted that some legumes are unpalatable to animals in the green condition (and have been rejected in plant evaluation experiments because of this) but are readily eaten as standing hay during the dry season. As noted elsewhere, this may aid plant survival (Section II-4) and may have the added advantage that it maximises N-fixation and provides high-quality feed in the dry season. Little or nothing is known of these plants, the factors that control their palatability or their feeding value.

Until recently, the main selection criteria for tropical legumes was that they persist and produce reasonable yields. This was sufficient when plant introduction programs were limited and few accessions met these requirements. It is now apparent that the tropics contains a wealth of such material which encompasses a wide variety of genera, species, and forms; in Stylosanthes alone there are now available annuals, biennials, and woody perennials all suited to dry tropical conditions. If such resources are to be sensibly developed that selection based upon animal require-ments is needed.

Nor does this requirement cease at this stage of plant devel-opment. The effects of species and management practices on animal reproductive rates need to be considered; indeed low reproductive rates are a major factor limiting animal production in the trop-ics. The same is true for lactation rates as these affect weaning weights and calf survival. The interactions between plant nutri-tion and occurrence of sub-clinical disease, which also greatly limit annual production, requires investigation. Finally, it is appropriate to mention the need for further work on supplementa-tion, the levels required for optimal animal production and the most appropriate source of the mineral to use. The utilization of 'low fertility' species may ultimately depend upon such knowledge.

By adding biologically fixed nitrogen to the farming system tropical pasture legumes can be used to increase animal produc-tion. To take full advantage of existing pasture legumes and to develop others and to fit them into farming systems throughout the tropical world we need to study plant-animal interactions.

REFERENCES

Playne, M.J. 1969. The nutritional value of intact whole pods of Townsville lucerne. Australian Journal of Experimental Agri-culture and Animal Husbandry 9: 50-57.
Preston, T.R. and Leng, R.A. 1978. Sugar cane as cattle feed. World Animal Review 27: 7-12.

5
Role of Legumes in Soil Improvement

I. Vallis, I. F. Fergus,
and E. F. Henzell

INTRODUCTION

Soil improvement will result from any process leading to changes in the chemical or physical properties of a soil, such that conditions become more favorable for plant growth. Clearly the degree and nature of the improvement possible will depend on the initial state of these properties. In the tropics, many (but by no means all) soils are highly weathered, acid, and of low chemical fertility, though they may have reasonable physical characteristics (e.g. oxisols and some alfisols). Others, particularly clay soils of the dry tropics, may exhibit physical limitations to plant growth (see Part III-1). The lighter textured topsoils of ultisols and some alfisols, often low in organic matter, have low water holding capacity and may possess hard-setting surfaces with poor rainfall acceptance which present serious erosion hazards.

In all these examples, some degree of soil improvement is often sought because the soils exhibit inherent features limiting to plant growth. Improvement may also be desired when soil productivity has declined following cropping. Maintenance of soil fertility and prevention of erosion are also important, especially in soils freshly cleared from forests. There is alarm over the extent of soil erosion in the tropics, principally caused by deforestation, departure from traditional shifting cultivation practices, and overgrazing of grasslands (Sanchez, 1976).

The role of legumes in effecting soil improvement depends primarily on two related processes--nitrogen (N) fixation and increases in soil organic matter. The objective may be (a) to maintain a permanent pasture system, (b) to develop a stable rotation which includes a cropping phase, or (c) to use the legume as an understory plant. In all cases, significant N-fixation depends on a satisfactory level of other nutrients and additions of fertilizers, particularly phosphorus (P), legume fertilizers are commonplace. Thus any discussion on the role of legumes in soil improvement should also be concerned with the combined effects of legume and fertilizer.

EFFECTS ON SOIL CHEMICAL PROPERTIES

Nitrogen and Organic Matter

In pastures that are grazed by beef cattle 90 percent or more of the N fixed by the legume is eventually added to the soil in plant litter and animal excreta. The amount and fate of this fixed N (e.g. accumulation in soil organic matter, availability for subsequent cycles of growth and losses from the system) will largely determine the effect of the legume on soil fertility.

The maintenance or increase of organic matter is important for the following reasons: it supplies most of the cation-exchange capacity (CEC) in acid, highly weathered soils; it reduces erosion and compaction hazards in sandy topsoils; it reduces P fixation by iron and aluminum oxides; it may form complexes with micronutrients which prevent their leaching; and it is an important slow-release source of N, S, and P for pasture or crop plants (Sanchez, 1976).

Within a specific environment N fixation by legumes has a close linear relation to dry matter production (Jones, 1972). Thus, given effective nodulation, factors that affect legume growth viz. soil moisture and nutrient status, solar radiation, competition from companion grasses, pests and diseases, and grazing management will also affect N fixation. The relationship between dry matter production and N fixation occurs because in grass-legume mixtures 80-95 percent of the legume N is usually derived from fixation (Vallis et al., 1977; Edmeades and Goh, 1978). However, the efficiency of N fixation may be reduced by high levels of soil mineral N when legume growth rates are low (Hoglund and Brock, 1978) or by high rates of N fertilizer (Ball et al., 1978).

We may note the view of Henzell (1968) that a guide to the range of N-fixation rates to be expected in practice is of more value than the results of specific experiments. By making some simple assumptions concerning the N content of legume tops and roots and the N cycle in pastures, Henzell (1968) estimated from fourteen sets of published data on legume dry matter production that N fixation by average growth of tropical legumes would be 20-180 kg ha^{-1} year^{-1}, with very good growth giving about 290 kg ha^{-1} year^{-1}. After numerous measurements of the dry matter production of tropical legumes during the subsequent decade there seems to be little reason to alter these estimates, though on rare occasions rates in excess of 300 kg ha^{-1} year^{-1} might occur in the humid tropics (Whitney et al., 1967; Grof and Harding, 1970). Furthermore, the range of results for Centrosema, Desmodium, and Stylosanthes are not markedly different, although the maximum for Stylosanthes is slightly higher. It must be stressed, however, that the above estimates are for well-fertilized swards with infrequent defoliation, where growth and N fixation would be expected to be greater than in commercially fertilized and grazed pastures.

The question of whether legume N fixation leads to an increase in soil N and organic matter and availability of N to plants is a complex matter, as changes in soil N content represent the balance between inputs and outputs. It is assumed that the input from symbiotic fixation is the most significant variable, and inputs from non-symbiotic fixation, fertilizers, or precipitation are insignificant or relatively constant. Whether factors that affect legume growth and symbiotic fixation have corresponding effects on accretion of soil N and organic matter will depend on how outputs (via volatilization of ammonia, leaching, runoff, denitrification, erosion, animal products, and fire) are affected. For example, if the extra legume growth in response to fertilizers is utilized by a higher stocking rate, then the extra input of fixed N would be balanced somewhat by greater volatilization of ammonia from excreta. Conversely, if a constant stocking rate is maintained while fixation is reduced by prolonged drought a measurable decrease in soil N can occur (Vallis, 1972).

Land preparation, fertilization, and grazing management are perhaps the most readily manipulated of the factors affecting the input of N by legumes. Well-prepared seedbeds with adequate fertilization would naturally be expected to give an early, rapid accumulation of soil N and organic matter (e.g. Crack, 1972). The possibility of N loss however, by erosion and leaching during land preparation should be kept in mind; early accumulation must compensate for these losses before there is any benefit (Myers, 1976). Observations with grazed pastures containing D. intortum and D. uncinatum show that the enhancement of N and organic matter accretion by application of superphosphate occurs only up to moderate rates of application (Henzell et al., 1966; Vallis, 1972; Bryan and Evans, 1973).

Nutrient availability may affect the incorporation of N into soil organic matter in another way besides through the effects on legume growth. Soil organic matter tends to have a more or less constant ratio of C:N:S:P[1] Williams and Donald, (1957), and Williams and Steinbergs (1958) suggested that organic matter accumulation could be limited by the amount of sulfur supplied in the superphosphate. Sulfur was also thought to be a factor limiting the accumulation of soil organic matter at higher rates of superphosphate application to subterranean clover on a sandy soil in Western Australia, but here loss by leaching of sulfur returned in sheep urine was implicated (Watson 1969). It is known that sulfur leaches readily from the surface of some deep sandy soils in the tropics (Gillman 1973), but the significance of this for accumulation of organic matter does not appear to have been studied.

The change in soil N following the establishment of a leguminous pasture is affected by the initial soil N content. This occurs even where there is no effect on botanical composition (Jones, 1967; Simpson et al., 1974). Two possible explanations for this are that at higher levels of soil N the efficiency of symbiotic fixation is lower (i.e. the legume gets more of its N from the soil and fixes less) or that losses of N are greater. Although there is evidence that high levels of soil mineral N or

---

[1]C:N:S:P = Carbon, Nitrogen, Sulfur, Phosphorus

fertilizer N can reduce the efficiency of N fixation by legumes in mixed pastures (Hoglund and Brock, 1978; Ball et al., 1978), there is also evidence that greater losses of N are a significant cause of lower rates of N accretion in soils of higher N content (Simpson et al., 1974). The level of soil N at which inputs and losses of N are in equilibrium can be surprisingly low. For example, Jones (1967) found no change under Macroptilium atropurpureum in soil with 0.15 percent N, compared with an accretion of 129 kg N ha^{-1} year^{-1} in a soil with a 0.09 percent N. At a higher level of soil N, Bruce (1965) found that under humid tropical conditions the inclusion of C. pubescens in pastures established on newly cleared land maintained the initial topsoil N content of 0.38 percent (in the absence of a legume soil N decreased by about 100 kg ha^{-1} year^{-1}. This can be compared with an annual increase of 280 kg N ha^{-1} in a topsoil initially having 0.07 percent N under a lightly grazed mixture of C. pubescens and giant stargrass (Moore, 1962).

A very common feature of N and organic matter accretion in soil under leguminous pastures is that the rate of accretion slows down as N accumulates, until eventually a point is approached where no further increase in soil N occurs. This feature is analogous to the effect of initial soil N content on the rate of N accretion, and the possible causes are the same, viz. increased competitive effects of the grass on legume growth, reduced efficiency of symbiotic fixation, and greater losses of N at higher levels of soil N. Although reduced efficiency of N fixation might be significant in virtually pure swards of legume (e.g. Wetselaar and Norman, 1960), ^{15}N-dilution measurements in a sequence of pastures of increasing age have shown that it is not likely to be important in mixed pastures (Edmeades and Goh, 1978).

The conclusion from the foregoing discussion is that although the general principles concerning the influence of a multiplicity of factors on the rate of soil N and organic matter build-up are understood, accurate predictions of the rates of change in new environments are not possible. So far we have considered only the amount of N accumulated by the soil-plant system. Equally important is its availabilty or rate of cycling in the system.

A major part of the increase in available N under legume pastures must come from decomposing legume residues and from animal excreta. During decomposition of plant residues, N is rapidly incorporated into microbial tissue and extra-cellular metabolites, with any N in excess of microbial needs being released as ammonium (Bartholomew, 1965; Oades and Ladd, 1977). During the decomposition some N may escape to the atmosphere as ammonia gas (Floate and Torrance, 1970). Subsequent release of N by decomposition of microbial cells and products is slower than that from plant residues, but faster than from the remainder of the soil organic matter. Over a period of several years, the availability of the added N approaches that of the bulk of the soil organic matter (Oades and Ladd, 1977).

Most of the information on the short-term mineralization of N in plant residues is from laboratory incubation of finely ground material. Very little information is available on the proportion of N in plant residues that becomes available for subsequent plant growth in the short term under field conditions, but it appears to be in the order of 15-30 percent in the first year (I. Vallis, unpublished data). This is considerably less than the proportion of dietary N excreted in the urine by cattle; Henzell and Ross, (1973) give a range of 33-73 percent, depending on the N concentration of the diet. Moreover, cycling via the animal takes only a day or two compared with weeks or months via residues. But it is doubtful whether this rapid and effective mineralization is of much advantage to the N economy of the pasture, because on sandy soils more than half of the urinary N can be lost as ammonia gas (Watson and Lapins, 1969). Large losses can also occur from faeces; Gillard (1967) estimated that 80 percent of the N in faecal pads was mineralized and subsequently lost as ammonia[2] but the loss was only 5-15 percent where dung-burying beetles were active. Furthermore, the distribution of excreta is uneven, with large quantities being deposited in unproductive areas where livestock drink and rest (Hilder, 1966). The relative importance of plant residues and excreta for the recycling of N will obviously depend on the degree of pasture utilization. In view of the importance attached to the fate of N excreted by grazing animals, remarkably few measurements have been made in tropical areas.

Other Nutrients

Accumulation of organic matter should lead to an increase in exchange capacity (for cations and anions), though the increase may only be significant in terms of soil improvement for soils with initially low exchange capacity. Such soils, either sandy ones or those with low activity clays, are common in the tropics (e.g. Uehara, 1978), and any increase in CEC (cation exchange capacity) would clearly be advantageous in enabling the soil to store larger amounts of cations. Whether such increases can be expected in practice is not clear. Williams and Donald (1957) found increases in CEC under temperate pastures, while Bryan and Evans (1973) measured increases under tropical pastures that included Desmodium. Lal et al. (1978) also observed increases under centro grown as a cover crop, but Talineau et al. (1976) reported an actual decrease in CEC under forage crops which included stylo and centro. Further, as was pointed out earlier, some studies of changes in soil properties under tropical legume pastures have shown no increase in soil organic matter (e.g. Meyers, 1976), and hence presumably, no increase in CEC.

Even if an increase in CEC is achieved, the content of plant nutrient cations in many tropical soils is so low that many of the new exchange sites are likely to be occupied by H^+. Soil improvement by additional storage of cations useful to plants may occur if deep-rooted legumes cycle such elements from lower horizons to the surface, though there seems no clear evidence of this for the three genera under discussion, or as a result of fertilizer addi-

---

[2]This is an unusually large loss of N that should be verified.

tions. The latter will only be significant where the amounts added are large, such as with lime or dolomite. This might be a desirable agronomic practice with soils of low base status, since there is evidence that stylo, desmodium, and centro all respond to lime in such situations (Munns et al., 1977; Trigoso and Fassbender, 1973). It should be noted though that in other cases desmodium responded but stylo did not (Dradu, 1974; Andrew, 1978).

Accumulation of organic matter in the presence of a legume carries the additional benefit that some other plant nutrients, as well as N, accumulate in a form from which they are gradually released by mineralization. Several factors must be considered before deciding whether this represents soil improvement. In the absence of fertilizer, the nutrients came originally from the soil and only indicate "improvement" if they are now in a form more useful to plants. This may be so if they are now less liable to loss by leaching, but the benefit would be more significant if the legume (due to mycorrhizal associations or other factors) had initially exploited forms not accessible to all plants. Such nutrients would eventually be returned via the organic cycle in forms more generally available to plants.

It is not clear whether this improvement is a likely event with soil P (phosphorus). Most published evidence suggests that plants do not differ in the pools of soil P they exploit (e.g. Nye and Foster, 1958; Marais et al., 1970), although the rates at which P is absorbed obviously vary. Russell (1978) points out that stylo has a lower P requirement (per g DM/day) than glycine, and centro a lower P requirement than desmodium. Plants with a lower requirement should grow better on P deficient soils; this does not necessarily mean they would give greater soil improvement. In fact, a low concentration of P in the herbage may slow down the rate of recycling of P. Critical factors may be the volume of soil explored by the root, or the limiting soil solution P concentration that the plant can utilize. Differences here could mean that one plant is potentially able to remove more soil P than another (Probert, 1978), particularly if P stress on soil reserves leads to mobilization of previously non-labile P by dissolution of sparingly soluble P compounds. Such a mechanism was proposed by Probert (1972) as a likely interpretation of his data. However, in recent experiments there were no differences between plant species in their ability to mobilize non-labile P (F.W. Smith, pers. comm.). Little information is available on possible differences in limiting soil solution P concentrations, although there is evidence of differences from flowing culture experiments (Asher and Loneragan, 1967). Exhaustive cropping would be one way of exploring this possibility, but there are some problems in experimental technique.

With low solubility P fertilizers, such as rock phosphate, there seems to be some possibility of selecting legumes with a superior ability to utilize the added P. For example, stylo and centro made better use of rock phosphate than did desmodium in experiments by Bryan and Andrew (1971). In practice, this could lead to greater soil improvement, using cheaper sources of P, with with some plants than with others. Although correction of P deficiency may be primarily aimed at increasing legume growth in the pasture, a further benefit in soil improvement could accrue in tropical soils with low activity clays, since correction of P deficiency may lower the point of zero charge, and so increase CEC, and possibly improve soil structure and increase water retention (Uehara, 1978).

Potassium (K) is likely to become a limiting nutrient in sustained legume growth in many tropical soils, particularly the high-weathered ones. In such cases the role of the legume in soil improvement will lie in increased CEC reducing leaching loss of K initially applied as fertilizer, and returned to the soil surface in a mobile form (as leaf leachate, in plant residues or in urine and, to a lesser extent, dung). In the latter case, the K will not be returned uniformly to the area, and may concentrate at sites where animals congregate (Hilder and Mottershead, 1963). For this reason, the efficiency of utilization of applied K in long-term pastures may be quite poor on a whole-pasture basis.

With less weathered soils, which have reserves of labile non-exchangeable K associated either with primary minerals or micacous clay minerals, the role of the legume may be more signifcant. Fergus and Martin (1974) presented evidence that plants differ in their ability to utilize K from non-exchangeable sources and showed that in some soils such uptake of initially nonexchangeable K commenced while plant K concentrations were still above the critical level. It may be possible, therefore, to use a legume with this capability to absorb K from forms not available to other plants in the pasture and to eventually return it to the soil surface in a form presumably available to all plants. Fergus and Martin (1974) showed that Macroptilium atropurpureum was superior to lucerne in its use of initially non-exchangeable K, and Gutteridge et al. (1976) found that desmodium did well on low-K soils. However, information is lacking on whether the legumes discussed differ in their use of less labile forms of K. Further, there may be some possibility of breeding cultivars which express this characteristic; Fergus and Hacker (unpublished) have shown that there are indeed intra-specific differences for Setaria.

The incorporation of sulfur (S) into organic matter is an integral part of the sulfur cycle in soils and an increase in the proportion of total sulfur that is in organic form may often lead to soil improvement which may occur if some plants utilize sorbed S better than others, or if deep-rooted plants tap reserves of S at depth, and so lead to surface soil enrichment; such deep reserves of S are a feature of some soils of high rainfall areas (Probert, 1977). A critical factor in the organic cycle will be the rate of mineralization of S. Clarke and Russell (1977) point out that this rate may be similar to the rate of release of N in temperate regions, but the aspect has not been studied extensively for tropical pastures. Reduced leaching losses when much of the S (possibly applied as a component of single superphosphate) is in organic form will be important in deep sandy soils, because Gillman (1973) has shown that sulphate is readily leached through such soils. But this will not be the case with many clay soils of the tropics, which have sufficient sorption sites to hold sulphate against leaching.

Trace elements will also be incorporated in the organic cycle, and legumes, because of their generally higher concentrations of such elements, may have a useful role. Once again there may be inter-specific differences in behavior; for example, Andrew and Thorne (1962) showed that desmodium had a lower requirement for copper than stylo. Elements such as copper, which is strongly complexed by soluble organic matter (Lindsay, 1978), and manganese and boron will be particularly involved in organic cycles. The role of organic matter in trace element availability under tropical pastures has not been extensively studied.

EFFECTS ON SOIL PHYSICAL PROPERTIES

Probably the most important physical property of soils for pasture use is their ability to store and supply water to plants. Water stress is far more likely to affect pasture production than is poor aeration; generally aeration is regarded as satisfactory for pastures if at least 10 percent of the soil volume is occupied by air when the soil is at field capacity (Isbell and McCown, 1977). The higher rainfall intensities in tropical areas compared with temperate areas mean that infiltration rates must also be higher or runoff and erosion hazards will be increased. There are large areas of alfisol soils used for cattle grazing of native and volunteer pastures in the tropics, at least in Australia, that have hard-setting A-horizons that shed water readily (Northcote et al., 1975). Unless water shed from the slopes is stored in the soils on lower land, less efficient use of rainfall will occur.

If the pasture is to be followed by an arable phase, the soil must additionally be able to withstand exposure to the action of rain and wind and the passage of machinery. That is, stability of structure is needed to avoid surface crusting, erosion, and compaction. Further, it needs to be easily tilled.

There are large areas of alfisols and ultisols in the tropics with sandy topsoils that are particularly prone to compaction, runoff, and erosion if they are exposed and cultivated (Sanchez, 1976). The best management for these soils is to maintain a protective plant cover or mulch, and one of the most effective plant covers is a pasture. Furthermore, an increase in soil organic matter, the mechanical effects of root growth, and increased activity of soil fauna can improve the physical properties of poorly structured soils. Bulk density can be reduced, and both porosity and stability of agggregation improved, leading to improved water entry and permeability and greater resistance to erosion (Greenland, 1971; Russell, 1971; Clark and Russell, 1977).

Two important questions are: will any improvement benefit the pasture itself or will it only be of benefit to a subsequent crop, and what specific role do legumes, especially Desmodium, Centrosema, and Stylosanthes spp., have? Benefits to the pasture would only be expected (but not ensured) if physical conditions at the start of pasture development were sub-optimal for pasture production and then only within the constraints set by soil texture and depth. The most likely situation is that of old arable land with degraded structure. In such cases the possibility that poor physical conditions might impair pasture establishment should be remembered (Roe, 1974).

The role of the legume seems to be simply that of increasing pasture growth and, in many cases, soil organic matter content. A good correlation between infiltration rate and previous amounts of growth (varied by the use of N fertilizer) has been shown for temperate grasses on a South Australian wheat-growing soil (Clarke, Greenland, and Quirk, 1967). Differences between grass species were also noted, Lolium rigidum being more effective than Phalaris tuberosa. Inter-specific differences between grasses have been ascribed to differences in root density and morphology (Clarke and Russell, 1977).

There is a paucity of data concerning the effects of tropical legumes on soil physical properties under pastures. Henzell et al. (1966) found that soil bulk density decreased under pastures containing D. uncinatum, but Crack (1972) and Vallis (1972) found no apparent trend in bulk density in pastures where soil N was accumulating. Elsewhere, improved soil porosity and permeability under S. guianensis and C. pubescens (Talineau et al., 1976) have been reported. Observations with temperate pastures indicate that effects on available water storage capacity are likely to be small (Russell and Shearer, 1964; Barrow, 1969) except where an organic 'mat' accumulates on the surface (Kleinig, 1966; Rixon and Bridge, 1967).

EFFECTS ON SOIL BIOLOGICAL CHARACTERISTICS

Increased biological activity, including roots, microorganisms and larger organisms would be expected to follow from increased pasture growth, when more organic matter is usually returned to the soil surface. As mentioned earlier, this activity would be anticipated when legume based pastures are grown to improve either already degraded soils or inherently unproductive ones. The main soil improvement due to enhanced biological activity will be shown in improved soil physical conditions and in an increased rate of nutrient cycling via organic matter. Although these effects have received considerable attention for temperate pastures, their significance under tropical pastures is far from clear.

For example, earthworms are commonly thought to be important under temperate pastures in improving aggregation and porosity (e.g. see Greenland, 1971). They are certainly more common under pasture than cropped soils (Russell, 1973). Stockdill (1966) found higher pasture yields after earthworms were introduced, and Sears and Evans (1953) observed a correlation between productivity of pasture and weight of earthworms. Other animals too will affect soil porosity and microaggregation (Burges and Raw, 1967). Clarke and Russell (1977) cautioned that the effects of soil fauna in improving soil aggregation may be too slow to have a great effect in short-term leys. However, earthworms are not as widespread in the tropics as in temperate areas (Lee 1974) and some species at least are intolerant of very acid soil conditions (< pH 4.5). Nevertheless, their importance or non-importance in the tropics, under permanent pastures in particular, warrants future study.

Termites and ants are important soil animals in the tropics, but there is little information on their role in decomposition of litter under grass/legume pasture. Lee and Wood (1971) pay close attention to the agricultural significance of termites, including their effects on organic matter decomposition, availability of nutrients, and soil structure. Although they cite examples of desirable changes resulting from at least some of these processes, they conclude that "in general, their effects are deleterious (to plant growth) and may lead to suppression of species that are favored as food, or, in extreme cases, to denudation of pastures."

There seems no doubt that considerably more information is needed on the role of soil animals under tropical pastures, particularly permanent pastures, before their significance in relation to soil improvement by legumes can be adequately assessed.

SUMMARY AND CONCLUSIONS

The role of the legume in soil improvement lies in:

(a) Increasing the N capital of the system, or the availability of N or both due to more rapid mineralization of organic forms which differ from those found in the absence of the legume.

(b) Increasing the organic matter content of the soil which affects a range of soil properties: (1) the OM acts as a store for nutrients, which are held either in organic combination or at charged surfaces; (2) in certain soils, e.g. alfisols and ultisols soil physical conditions are improved. Bulk density is lowered, aggregation improved, and pore space, available water storage, and infiltration rate increased and susceptibility to erosion reduced; (3) increasing biological activity which affects a more rapid cycling of nutrients in the organic pool, and contributes to the improvement in aggregation and porosity. The relative importance of these effects will clearly vary from soil to soil.

(c) Providing a more effective vegetative cover in those situations where growth of pasture or cover crop is restricted by N deficiency, thus reducing the amount of bare soil exposed to erosive forces.

(d) Effecting, indirectly, the soil because fertilizer nutrients often added to achieve acceptable legume growth will themselves lead to soil improvement. If N is deficient, the advantage of the legume is in rapid uptake of applied nutrients, reducing losses due to leaching or sorption processes. If N is adequate, competition between grass and legume may mean that it is the pasture as a whole that fills this role. If particular species or strains are able to use nutrients not accessible to other plants, the amount of generally available nutrients will be increased.

Grazing pressure will markedly modify all of the foregoing, since heavier stocking will reduce plant cover and alter the ways in which nutrients are cycled; the rate of release of nutrients via dung and urine will be more rapid, but the distribution of cycled nutrients will be much less uniform and losses (especially of N) may be greater.

FUTURE RESEARCH NEEDS

We can state that further research aimed at a more complete understanding of any land-use system will always be necessary, though the benefits from such studies on already highly developed systems may be marginal. Tropical pastures cannot at present be regarded as highly developed systems, so benefits from a wide range of further research should be substantial; nevertheless, in the following we have attempted to focus on those areas where we see the greatest deficiencies in need at the moment, with respect to soil improvement:

(a) Further information is needed on the rate of organic matter build-up under the three legumes being considered and on their effect on the N nutrition of associated or subsequent crops. Extend present knowledge to more soil and climatic environments; the relevant environmental information should be thoroughly documented in reporting on such studies.

(b) Further information is needed on the rates of nutrient cycling in pasture ecosystems particularly the rates of mineralization of N, P, and S, because organic matter accumulation is of limited benefit if recycling of elements is slow. Release of P from organic forms may well be slower than release of N and S (Greenland, 1971; Clarke and Russell, 1977). Hence there is a particular need for work on rates of P release, for long term immobilization of P in organic matter is obviously undersirable.

(c) Better understanding of the role of ley farming in tropical agricultural systems, and in particular the significance of legumes in short-term rotations, is not well understood.

(d) Further investigate of whether there is a significant increase in CEC in soils with low activity clays under long term pastures.

(e) Better quantify changes in soil physical conditions under tropical pastures. Measurements of infiltration rate, and amounts of available water stored, are particularly relevant for for soils characterized by hard-setting surfaces, which shed high intensity rain readily. Such soils are common in the semi-arid tropics. It is here that improvement in soil physical conditions is most desirable.

(f) Study in detail to clarify whether the legumes under consideration differ in the amounts of P they can extract either from soil reserves or, perhaps more importantly, from less soluble P fertilizers. Similar information is required for other nutrients too - particularly K and S.

(g) Quantify data on the extent to which the return of nutrients in grazed pastures is concentrated at sites favored by animals. Clearly such concentrations do not represent soil improvement in the pasture as a whole.

(h) Study changes in biological activity, particularly under permanent pasture. Emphasizing the activities of termites, ants, and earthworms but not exclusively so. Correlations should be sought between biological activity and changes in soil physical and chemical properties.

REFERENCES

Andrew, C.S. 1978. Mineral characterization of tropical forage legumes. In "Mineral Nutrition of legumes in Tropical and Subtropical Soils" (eds.) C.S. Andrew and E.J. Kamprath. C.S.I.R.O. Melbourne. pp. 93-112.

Andrew, C.S., and Thorne, P.M. 1962. Comparative responses to copper of some tropical and temperate pasture legumes. Australian Journal of Agricultural Research 13: 821-835.

Asher, C.J., and Loneragan, J.F. 1967. Response of plants to phosphate concentration in solution culture. I. Growth and phosphorus content. Soil Science 103: 225-233.

Ball, R., Molloy, L.F., and Ross, D.J. 1978. Influence of fertilizer nitrogen on herbage dry matter and nitrogen yields, and botanical composition of a grazed grass-clover pasture. New Zealand Journal of Agricultural Research 21: 47-55.

Barrow, N.J. 1969. The accumulation of soil organic matter under pasture and its effect on soil properties. Australian Journal of Experimental Agriculture and Animal Husbandry 9: 437-444.

Bartholomew, W.V. 1965. Mineralization and immobilization of nitrogen in the decomposition of plant and animal residues. In "Soil Nitrogen" (eds.) W.V. Bartholomew and F.E. Clark, American Society of Agronomy, Madison, U.S.A.

Bruce, R.C. 1965. Effect of Centrosema pubescens Benth. on soil fertility in the humid tropics. Queensland Journal of Agriculture and Animal Sciences 22: 221-226.

Bryan, W.W., and Andrew, C.S. 1971. The value of Nauru rock phosphate as a source of phosphorus for some tropical pasture legumes. Australian Journal of Experimental Agriculture and Animal Husbandry 11: 532-535.

Bryan, W.W., and Evans, T.R. 1973. Effects of soils, fertilizers and stocking rates on pastures and beef production on the Wallum of south east Queensland. I. Botanical composition and chemical effects on plants and soils. Australian Journal of Experimental Agriculture and Animal Husbandry 13: 516-529.

Burges, A. and Raw, F. 1967. "Soil Biology". Academic Press, London.

Clarke, A.L., Greenland, D.J., and Quirk, J.P. 1967. Changes in some physical properties of the surface of an impoverished red-brown earth under pasture. Australian Journal of Soil Research 5: 59-68.

Clarke, A.L. and Russell, J.S. 1977. Crop sequential practices. In "Soil Factors in Crop Production in a Semi-arid Environment'. (eds.) J.S. Russell and E.L. Greacen. University of Queensland Press, St. Lucia.

Crack, B.J. 1972. Changes in soil nitrogen following different establishment procedures for Townsville stylo on a solodic soil in north-eastern Queensland. Australian Journal of Experimental Agriculture and Animal Husbandry 12: 274-280.

Dradu, E.A.A. 1974. Soil fertility studies on loam soils for pasture development in Uganda: Pot experiments. East African Agriculture and Forestry Journal 40: 125-131.

Edmeades, D.C. and Goh, K.M. 1978. Symbiotic nitrogen fixation in a sequence of pastures of increasing age measured by N dilution technique. New Zealand Journal of Agricultural Research 21: 623-628.

Fergus, I.F. and Martin, A.E. 1974. Studies on potassium. IV. Interspecific differences in the uptake of non-exchangeable potassium. Australian Journal of Soil Research 12: 147-158.

Floate, M.J.S. and Torrance, C.J.W. 1970. Decomposition of the organic materials from hill soils and pastures. I. Incubation method for studying the mineralisation of carbon, nitrogen and phosphorus. Journal of Science for Food and Agriculture 21: 116-120.

Gillard, P. 1967. Coprophagous beetles in pasture ecosystems. Journal of the Australian Institute of Agricultural Science 33: 30-34.

Gillman, G.P. 1973. Studies on some deep sandy soils in Cape York Peninsula, North Queensland. III. Losses of applied phosphorus and sulphur. Australian Journal of Experimental Agriculture and Animal Husbandry 13: 418-422.

Greenland, D.J. 1971. Changes in the nitrogen status and physical condition of soils under pastures, with special reference to the maintenance of the fertility of Australian soils used for growing wheat. Soils and Fertilizers 34: 237-251.

Grof, B. and Harding, W.A.T. 1970. Yield attributes of some species and ecotypes of Centrosema in North Queensland. Queensland Journal of Agriculture and Animal Sciences 27: 237-243.

Gutteridge, R.C., Whiteman, P.C., and Watson, S.E. 1976. "Final report on the regional programme of pasture species evaluation and soil fertility assessment, 1973-1976." University of Queensland Press, St. Lucia.

Henzell, E.F. 1968. Sources of nitrogen for Queensland pastures. Tropical Grasslands 2: 1-17.

Henzell. E.F., Fergus, I.F., and Martin A.E. 1966. Accumulation of soil nitrogen and carbon under a Desmodium uncinatum pasture. Australian Journal of Experimental Agriculture and Anmal Husbandry 6: 157-160.

Henzell, E.F. and Ross, P.J. 1973. The nitrogen cycle of pasture ecosystems. In "Chemistry and Biochemistry of Herbage". Vol. 2. (eds.) G.W. Butler and R.W. Bailey. Academic Press, London.

Hilder, E.J. 1966. Distribution of excreta by sheep at pasture. Proceedings 10th International Grassland Congress, Helsinki, Finland. pp. 977-981.

Hilder, E.J. and Mottershead, B.E. 1963. The re-distribution of plant nutrients through free-grazing sheep. Australian Journal of Science 26: 88-89.

Hoglund, J.H. and Brock, J.L. 1978. Regulation of nitrogen fixation in a grazed pasture. New Zealand Journal of Agricultural Research 21: 73-82.

Isbell, R.F. and McCown, R.L. 1977. Land. In 'Tropical Pasture Research Principles and Methods' (eds.) N.H. Shaw and W.W. Bryan. Commonwealth Bureau of Pastures and Field Crops (U.K.), Hurley. Bull. 51.

Jones, R.J. 1967. The effects of some grazed tropical grass-legume mixtures and nitrogen fertilized grass on total soil nitrogen, organic carbon and subsequent yields of Sorghum vulgaris. Australian Journal of Experimental Agriculture and Animal Husbandry 7: 66-71.

Jones, R.J. 1972. The place of legumes in tropical pastures. ASPAC Tech. Bull. 9 Taipei, Taiwan.

Kleinig, C.R. 1966. Mats of unincorporated organic matter under irrigated pastures. Australian Journal of Agricultural Research 17: 323-333.

Lal, R., Wilson, G.R., and Okigbo, B.N. 1978. No-till farming after various grasses and leguminous cover crops in tropical Alfisol. I. Crop performance. Field Crops Research 1: 71-84.

Lee, K.E., and Wood, T.G. 1971. "Termites and Soils." Academic Press, London.

Lee, K.E. 1974. The significance of soil animals in organic matter decomposition and mineral cycling in tropical forest and savanna ecosystems. In "Trans. 10th International Congress of Soil Science" 3: 43-50.

Lindsay, W.L. 1978. Chemical reactions affecting the availability of micro-nutrients in soils. In "Mineral Nutrition of Legumes in Tropical and Subtropical Soils." (eds.) C.S. Andrew and E.J. Kamprath. C.S.I.R.O. Melbourne. pp. 153-168.

Marais, P.G., Deist, J., Harry, R.B.A., and Heyus, C.F.G. 1970. Ability of different plant species to absorb phosphate. Agrochemophysica 2: 7-12.

Moore, A.W. 1962. The influence of a legume on soil fertility under a grazed tropical pasture. Empire Journal of Experimental Agriculture (U.K.) 30: 239-248.

Munns, D.N., Fox, R.L., and Koch, B.L. 1977. Influence of lime on nitrogen fixation by tropical and temperate legumes. Plant and Soil 46: 591-601.

Myers, R.J.K. 1976. Nitrogen accretion and other soil changes in Tindall clay loam under Townsville stylo/grass pastures. Australian Journal of Experimental Agriculture and Animal Husbandry 16: 94-98.

Northcote, K.H., Hubble, G.D., Isbell, R.F., Thompson, C.H., and Bettenay, E. 1975. "A Description of Australian Soils". CSIRO, Melbourne, Australia.

Nye, P.H. and Foster, W.N.M. 1958. A study of the mechanism of soil-phosphate uptake in relation to plant species. Plant and Soil 9: 338-352.

Oades, J.M. and Ladd, J.N. 1977. Biochemical properties: carbon and nitrogen metabolism. In "Soil Factors in Crop Production in a Semi-Arid Environment" (eds.) J.S. Russell and E.L. Greacen. University of Queensland Press, St. Lucia.

Probert, M.E. 1972. Dependence of isotopically exchangeable phosphate (L-value) on phosphate uptake. Plant and Soil 36: 141-148.

Probert, M.E. 1977. The distribution of sulphur and carbon-nitrogen-sulphur relationships in some north Queensland soils. CSIRO Australia Division of Soils, Technical Paper No. 31.

Probert, M.E. 1978. Availability of phosphorus and sulphur in tropical soils in relation to legume growth. In 'Mineral Nutrition of Legumes in Tropical and Subtropical Soils' (eds.) C.S. Andrew and E.J. Kamprath. C.S.I.R.O. Melbourne, Australia. pp. 347-360.

Rixon, A.J., and Bridge, B.J. 1967. Soil fertility changes in a red-brown earth under irrigated pastures. III. Water relations of the mat and surface soil. Australian Journal of Agricultural Research 18: 741-753.

Russell, E.W. 1971. Soil structure: its maintenance and improvement. Journal of Soil Science 22: 137-151. (U.K.)

Russell, E.W. 1973. "Soil Conditions and Plant Growth" 10th Ed. Longman, London.

Russell, J.S. 1978. Soil factors affecting the growth of legumes on low fertility soils in the tropics and subtropics. In "Mineral Nutrition of Legumes in Tropical and Subtropical Soils." (eds.) C.S. Andrew and E.J. Kamprath. C.S.I.R.O. Melbourne, Australia. pp. 75-92.

Russell, J.S. and Shearer, R.C. 1964. Soil fertility changes in the long-term experimental plots at Kybybolite, South Australia. IV. Changes in certain moisture characteristics. Australian Journal of Agricultural Research 15: 91-100.

Sanchez, P.A. 1976. "Properties and Management of Soils in the Tropics" John Wiley and Sons, New York.

Sears, P.D. and Evans, L.T. 1953. Pasture growth and soil fertility. III. The influence of red and white clovers, superphosphate, lime and dung and urine on soil composition, and on earth-worm and grass-grub populations. New Zealand Journal of Science and Technology 35A, Suppl. 1, 42-52.

Simpson, J.R., Bromfield, S.M., and Jones, O.L. 1974. Effects of management on soil fertility under pasture. 3. Changes in total soil nitrogen, carbon, phosphorus, and exchangeable cations. Australian Journal of Experimental Agriculture and Animal Husbandry 14: 487-494.

Stockdill, S.M.J. 1966. The effect of earthworms on pastures. Proceedings of New Zealand Ecological Society 13: 68-75.

Talineau, J.C., Hainnaus, G., Bonzon, B., Eillornneau, C., Picard, D., and Sicot, M. 1976. Agronomic aspects of the inclusion of a forage crop in a crop rotation under humid tropical conditions in the Ivory Coast. Cahiens ORSTROM, Biologie 11: 277-290.

Trigoso, R. and Fassbender, H.W. 1973. Effect of applications of Ca, Mg, P, Mo and B on yield and nitrogen fixation in some tropical legumes. Turrialba 23: 172-180.

Uehara, G. 1978. Mineralogy of the predominant soils in tropical and subtropical regions. In "Mineral Nutrition of Legumes in Tropical and Subtropical Soils" (eds.) C.S. Andrew and E.J. Kamprath. C.S.I.R.O., Melbourne. pp. 21-36.

Vallis, I., 1972. Soil nitrogen changes under continuously grazed legume-grass pastures in subtropical coastal Queensland. Australian Journal of Experimental Agriculture and Animal Husbandry 12: 495-501.

Vallis, I., Henzell, E.F., and Evans, T.R. 1977. Uptake of soil nitrogen by legumes in mixed swards. Australian Journal of Agricultural Research 28: 413-425.

Watson, E.R. 1969. The influence of subterranean pastures on soil fertility III. The effect of applied phosphorus and sulphur. Australian Journal of Agricultural Research 20: 447-456.

Watson, E.R. and Lapins, R. 1969. Losses of nitrogen from urine on soils from south-western Australia. Australian Journal of Experimental Agriculture and Animal Husbandry 9: 85-91.

Wetselaar, R. and Norman, M.J.T. 1960. Soil and crop nitrogen at Katherine, N.T. CSIRO, Australia, Division of Land Research and Regional Survey Technical Paper No. 10.

Whitney, A.S., Kanehiro, Y., and Sherman, G.D. 1967. Nitrogen relationships of three tropical forage legumes in pure stands and in grass mixtures. Agronomy Journal 59: 47-50.

Williams, C.H. and Donald, C.M. 1957. Changes in organic matter and pH in a podzolic soil as influenced by subterranean clover and superphosphate. Australian Journal of Agricultural Research 8: 179-189.

Williams, C.H. and Steinbergs, A. 1958. Sulphur and phosphorus in some eastern Australian soils. Australian Journal Agricultural Research 9: 483-491.

6
Social and Economic Considerations

D. G. Cameron
R. L. Burt

INTRODUCTION

Pastures are planted or not planted for many reasons; most are economic but social considerations are also important. Similarly, not all sown pastures are legume based and the reasons for this again vary. In the brigalow and gidgee lands of Queensland for example, the soils are quite fertile and no well adapted legumes are available. On clearing, extensive pure grass pastures are planted. On the other hand, under intensive production systems, grass/bag nitrogen pastures are often used to obtain maximum production per unit area. Generally, however, we are dealing with grass/legume pastures, the grass being sown or native.

In a previous chapter (Section I-6, 7) we have had two contrasting examples of pasture development. In these areas legume based pastures have been successfully developed and used to economically increase beef production. The incentives, methods, plants, and management systems used were quite different. Each, however, was appropriate to the climate, economic, and social system of the areas involved. Had they not been so, the enterprises would almost certainly have failed. In our examples, the two approaches are not interchangeable even though the areas are geographically, though not climatically, very close and are producing for the same markets.

These examples stress the need to have clear objectives when developing any tropical pasture enterprise. Many such projects have failed because the objectives were not clearly seen, and incorrect species and development methods were used. Plants from dryland extensive systems will rarely suit an intensively managed area with higher rainfall. Similarly, the development techniques from intensive areas do not suit extensive pastures in drier country. The costs involved by these intensive techniques are beyond those recoverable from a pasture in dry country where low stocking rates must be used.

In this essay, we will briefly discuss some of the social and economic considerations involved in pasture development. Most of the information and examples of necessity are from Australian situations - this is understandable--the concept of improved pastures in the tropics is relatively new and many varieties suitable for use in some areas have only recently become available. We will, however, refer to work elsewhere and try to highlight some of the problems involved.

THE PURPOSE OF LEGUME-BASED PASTURES

The purpose for which a pasture is to be used will play a major role in determining the type of pasture and the manner in which it is developed and managed. The incentives for development also play a role in determining the type of pasture planted.

Improvement in animal production may not be the only or even the main reason for sowing pastures. In Southern Australia subterranean clover and annual medic-based pastures are widely sown as a soil fertility building ley rotation with wheat. The species involved are hard seeded annuals well suited to the task. The seed is able to survive in the soil over cropping periods, and subsequently regenerates the pasture. Because of the resultant benefits to the following wheat crops, these pasture periods continue to be used at times when livestock industries are depressed.

Somewhat similar conditions occur in many tropical areas and the cropping systems there would benefit from legume based ley rotations. In Thailand and Laos, University of Queensland researchers are studying, in cooperative programs, the introduction of Stylosanthes guianensis into upland rice areas. This can be done without adversely affecting the yields of rice grain (Shelton and Humphreys, 1972) especially if the stylo planting is delayed slightly. The quantity and quality of the dry season grazing from the paddy lands after the rice harvest is then improved, an important feature in livestock production in that area. These techniques have been successful at both Na Pheng in Laos (Thomas and Humphreys, 1970) and Khon Kaen in Thailand (Shelton and Humphreys, 1974). As well as providing an addition to the dry season forage supply, upland rice areas so planted have had less weeds in the following crop, less erosion and some possibilites of increased soil fertility.

Pastures are also often planted, especially by absentee land owners, as a status symbol or as a means of reducing income tax payments rather than to directly improve the productivity of their land. In these cases, people whose major income is from other sources frequently use funds generated by the main income source to develop "model" farms; for prestige reasons, high quality pastures are demanded.

The intensity of production aimed at and permitted by the local environment will also determine the type of pasture planted. In areas where only extensive production is possible large areas have to be planted; the standards demanded both in the development techniques and the resultant pastures will be much different from those in areas where intensive development is possible. Not only the climate and soils of the area but also the size of the properties will at times determine whether extensive or intensive pasture development occurs.

In extensive areas it costs substantial amounts just to purchase the seed and have it distributed with minimum fertilizer input. As a result, plants that do not require sophisticated seed beds are necessary. Often, all that is involved is the aerial seeding of legumes into the native grasses. At its lowest level, this may achieve little more than carrying the same number of stock on a better plane of nutrition. Productivity from such pasture species is not the only consideration; rather it is ease of establishment and ability to naturalize into the dominant environment and to improve the level of nutrition at critical times of the year.

A different situation may exist on small properties; where it may be necessary to intensify the level of production to somewhere near its biological maximum if the enterprise is to remain a viable economic proposition. Fully cleared land and well cultivated seed beds will be required and productivity, and persistence under heavy grazing, rather than ease of establishment, will determine the species to be used.

Thus, the purpose of the pasture will clearly define the type of plant to be used. The appropriate one may not exist in currently available commerical cultivars but may be among those discarded during their selection for their particular niches. As a result, old collections will need to be re-evaluated with the new requirements in mind.

Probably the first requirement is to determine whether improved pastures are needed at all and this decision is particularly relevant in extensive production systems. If extensive areas of land are available it is often more economical to contemplate for instance, the acquisition of more land; more stock can be carried without much being spent on development per se. This can also have social as well as economic incentives where land ownership is a major status symbol. In areas capable of high production on the other hand the natural grazing is probably being over stressed. Here the main need will be for much higher levels of dry matter production to give a viable enterprise. If these levels need to be extremely high then it may be necessary to omit the legume as a pasture base and nitrogen source and go directly to pure grass pastures fertilized with bag nitrogen.

Sown pastures in more favored situations on larger properties enable more effective control of stock. By concentrating them on small more closely managed areas labor costs are reduced, important in high wage situations. Similarly, in less productive areas, only store cattle can normally be turned off. The establishment of adequate areas of legume based pastures in these areas permits the cattle to be finished to slaughter condition on the property.

Pastures can also be required for other reasons, especially on larger properties. On these the only plantings may be in the bull paddock to ensure the bulls go out each year in good condition, in the horse paddock to reduce feed bills, or as a hospital paddock, for sick or injured animals, to speed their recovery.

Many other uses of pasture are also being contemplated at present. In Antigua, in the Caribbean, waste sludge from poultry and pigs is spread onto a mixture of grasses, legumes, and sugar cane, and the feed for the animals concerned is fitted into a cut-and carry system. This helps to reduce the need to feed costly imported concentrates in the dry season (Keoghan, pers. comm. 1979). In Thailand, Verano stylo (Stylosanthes hamata) is sown onto the roadsides and the resultant pastureage used to feed working oxen. In other areas the pelleting of the excess wet season legume production for chicken feed is being considered. Such developments are new and their success or failure is dependent on complex socio-economic conditions. All, however, are dependent upon the provision of well adapted legumes capable of providing biologically fixed nitrogen.

INCENTIVES FOR PASTURE DEVELOPMENT

On properties that are sufficiently large to provide an adequate living for the landholder in the undeveloped or partially developed stage there may be no great incentive to develop their productive potential further. If for community reasons, higher levels of production are required from such land, then it may be necessary to use legislative processes to reduce property size and further stimulate pasture development and hence productivity. On large holdings, where the planting of improved pasture has not yet become a matter of habit and the landholder is already making a reasonable living, sown pastures represent an unknown higher risk factor which the grazier often prefers to avoid. Sown pastures take on slowly in such areas.

It is on smaller holdings, where the incentive to fully utilize a sown pasture technology is greatest, that such sowings expand most rapidly. The major restraint here is often ability to borrow or generate the necessary finance to undertake the development and here the provision of appropriate low interest public funds can be the greatest incentive to speeding development.

Not only do the landholders have to be convinced of the profitability of pasture improvement before it will be widely accepted in an area but it is also important that the people who service rural industries be convinced. This particularly applies to bank managers and other sources of finance such as stock and station agents and directorates of major pastoral houses.

All these factors mean that where an adequate technology for widespread pasture development has recently evolved it will not be widely taken up overnight. Adaptation will be quite a slow process gaining momentum only as confidence grows in its usefulness and profitability. It is important therefore that no mistakes are made in the development of the technology and particularly that no premature releases of information to the landholders occur. Should the initial innovators suffer reverses due to faulty information or inadequately tested technology then the final development of larger areas of sown pasture will be seriously delayed.

ORDERS OF DEVELOPMENT

Where landholdings are large there are several development approaches that can be followed. That which often operated in South Eastern Queensland was to first ensure that all sections of the property were adequately watered so that the livestock could readily graze all available land. Secondly, it ensured that at least basic fencing was erected to provide scope for livestock management and reliable segregation of differing classes of stock. When these two stages were complete the landholders then proceeded with removal of unnecessary timber species and woody weeds, leaving adequate timber for shade and future fencing materials, permitting much greater native grass production and therefore higher carrying capacity. It was also considered necessary before subsequent pasture development was attempted.

It is now known, however, that where timber is sparse and of an open savannah type, it may be possible to move directly into a pasture development phase upgrading the natural grasslands with introduced legumes to improve the nutritive value of the feed produced at critical times of the year or at least to narrow the gap between periods of quality feed.

INTENSITY OF DEVELOPMENT

As already emphasized, levels of intensity of land use will vary markedly depending on climate, soil type, and property sizes. Not only will the final level of development be determined by the relative costs of such development and the expected returns modified by various social factors, but it is also important to compare the relative risk involved. This is especially so in areas of erratic rainfall where the risk is less if large areas are developed to a lower level of overall productivity than if smaller areas are intensively developed. Intensive development is costly and such areas must be heavily utilized to recoup the development cost. As a result they are operating nearer to the crash point and soon both the stock and pasture succumb to drought stresses. Extensive areas are rarely, and certainly not wisely, operated so near to these limits, and so more slack is available in the system to absorb stresses due to drought or fire.

In erratic rainfall and drought prone country the larger holding has an advantage in escaping drought effects. There is always the prospect of isolated storms falling on one end of a large property so that stock can be concentrated within the favored area. Following the storms is only possible with smaller holdings by using adjustment processes and is much more difficult to achieve.

THE COSTS IN A PASTURE SYSTEM

The costs in a pasture system are diverse and can be considerable, especially when a heavy body of timber has to be removed to plant the pasture. The basic costs are for seed and fertilizer, both purchase and spreading, land preparation, and further fencing and watering.

Seed costs are a function of the amount produced per unit area and its ease of harvesting, handling, and cleaning. They are usually lessened as a cultivar becomes more widely used. At this stage it is produced as a sideline on general grazing areas, whereas in the early stages specialist seed producers handle its production. The perfection required in the seed bed influences its cost. Weed free, fine, firm seed beds are expensive, but some very productive plants do not readily tolerate shoddy, ill prepared seedbeds. On the other hand plants for extensive pastures need to naturalize from sparse initial establishments in a rudimentary seedbed. Similarly seeding rate and so total seed cost is a function of the establishment potential of the pasture components and the quality and density required in the seedling stand.

Fertilizer costs revolve around both establishment and maintenance applications and the rates and types required. They are essentially related to soil type but it is being found that Stylosanthes spp. can be grown with less than normal phosphatic fertilizer applications.

The degree of initial land development is also an important cost. It is much cheaper to establish pastures on old crop land or open grazing areas, providing there is no weed problem, than is the case if the original timber has to be removed first. The size and density of this timber also influences the development cost. Dense stands of fine-stemmed timber can often be burnt as they lay on the ground following pulling. This is the case with brigalow and gidgee, whereas in most eucalypt forests, it is necessary to burn the timber in windrows and to rework these periodically during burning to get complete burning of all sticks and logs.

The total area of pasture available at a given time and that potentially available will often determine how an area is used. Often little impact on productivity is seen until a minimal area per livestock unit is available. On the dairy farms of the near north coast of southeast Queensland there appears to be a threshold area of about one hectare per cow before sown legume pastures become effective and stable. The viny legumes, Siratro (Macroptilium atropurpureum) and Greenleaf desmodium (Desmodium intortum) are unstable and susceptible to overgrazing in stress periods below this threshold.

The selected management system to be used also affects the cost of pastures. A sophisticated rotational grazing system requires much greater expenditure of fencing and watering than does the large set stocked paddock normally used under extensive systems.

Distance from service centers and the quality of the transport system is also important. If seed and fertilizer brought onto the property and stock for market have to be transported long distances over bad roads, substantial differences are made to profitabilities. Robinson and Sing (1975) calculated that in North Queensland, considering the then existing prices (1975) and fertilizer recommendations for Townsville stylo (Stylosanthes humilis) about 400 km from the port of Cairns was the maximum likely distance at which it would be profitable to develop pastures based on Townsville stylo. Upgrading the road system in the Cape York Peninsula area to "Beef Road" standard, i.e. the main roads to a sealed road, would extend this profitability to 700 km by road from Cairns, other costs remaining the same. This extra 300 km radius could have lead to more favorable development on an extra 1/2 million km of land or an area twice as large as the United Kingdom. Reducing the amount of maintenance fertilizer by 25 percent with the then existing (1975) road system increased the limits of profitability from Cairns by 160 km to about 550 km.

FERTILIZER COSTS

On less fertile soils the major long term cost in the use of sown pastures can be the fertilizer costs. If we are to ensure adequate legume and pasture growth and healthy animals which breed and grow well, there must be adequate levels of minerals in the ecosystem. This can be achieved in several ways. Fertilizer can be added to the soil and from there to the plants and animals, or the minerals can be provided directly to the animals as supplement.

In extensive situations it may be more economic to seek plants with lower mineral requirements for growth, Stylosanthes low phosphate requirements being an example, and supplement the animals with the deficient minerals. It might even be effective to have the plants operating at suboptimum levels over large areas than growing well over smaller areas. Most early fertilizer research was aimed at determining levels required for optimum growth. Over recent years the emphasis has tended to be on low input fertilizer systems, but much has yet to be done before all the answers are known.

In Australia, at least, nitrogen is probably the most expensive nutrient to apply and is often the one most widely limiting pasture performance. We have seen how legumes can be used to provide this nitrogen and at the same time, themselves provide valuable animal feed. In very intensive production systems however, it may be necessary to go to a grass/bag nitrogen system to get maximum animal product per unit area.

That legumes can however, make a substantial contribution to the nitrogen supply in farming systems is illustrated by Donald (1960) who estimated that sown pastures in southern Australia were adding in the order of 1,000,000 tons of nitrogen per annum to the soil. By 1970, Williams and Andrew (1970) estimated this to be 1,500,000 tons due mainly to increased areas planted and worth, at this stage, $300,000,000 in equivalent nitrogen fertilizers. These figures, and especially the value, would be substantially expanded with the steep increases of recent years in nitrogen fertilizer prices.

There are many nitrogen deficient soils in the tropics. The "cerrados" of South America represent one such type. These soils occupy an area of just under 2 million km and to use them intensively for agriculture would require applications of about 100 kg nitrogen ha^{-1} annum^{-1}, requiring about 40 percent of the world's current output of nitrogenous fertilizer (Dobereiner, 1978). It seems likely that adapted pasture legumes can be found for such situations; some with yields over 15 t ha^{-1} annum^{-1}. Clearly, therefore, the search for adapted legumes and the relevant technology to go with them is vitally important.

As with other fertilizers nitrogen fertilized grass pastures may have a role to play at suboptimum levels of fertilizer application and as only part of an overall system. Teitzel et al. (1974) advocated a stress relief system for the legume/grass pastures on the wet tropical coast of North Queensland in which one quarter of the sown pasture area is stoloniferous grass/bag nitrogen and the remainder legume/grass and the nitrogen is only applied during winter/spring seasons and the grass areas used to take extra grazing when legume growth is slow.

CONCLUSIONS

It is difficult to sensibly project finance into many tropical situations because lack of understanding of all the implications of such actions. The use of adapted pasture legumes in a farming system is one method of injecting biologically fixed nitrogen into that system. Such a nitrogen input can result in increased animal production and can be used to sustain crop growth. Its monetary value can be extremely large.

It seems that such an approach would be of particular value in developing countries; certainly once the relevant legumes have been domesticated, it would produce a cheap, long lasting beneficial effect. It is, however, extremely difficult to transfer technology from one area to another or even to choose objectives which are socially and economically acceptable by the recipient area. In this section we have done little more than highlight some of the more practical considerations inherent in pasture development.

REFERENCES

Dobereiner, J. 1978. Potential for nitrogen fixation in tropical legumes and grasses. In "Limitations and Potentials for Biological Nitrogen Fixation in the Tropics." (eds.) J. Dobereiner, R.H. Burrin, A. Hollaender, A.A. Francs, C.A. Neyia, and D.B. Scott. Plenum Press, New York and London. pp. 13-24.

Donald, C.M. 1960. The impact of cheap nitrogen. Journal of the Australian Institute Agricultural Science 26: 319-338.

Robinson, I.B. and Sing, N.C. 1975. Estimated profitability of investment in legume pasture development in far north Queensland. Tropical Grasslands 9: 199-207.

Shelton, H.M. and Humphreys, L.F. 1972. Pasture Establishment in Upland Rice Crops at Na Pheng, Central Laos. Tropical Grasslands 6: 223-228.

Shelton, H.M. and Humphreys, L.R. 1974. Factors influencing competition between upland rice (Oryza sativa) and undersown stylo (Stylosanthes guianensis) Working Papers XII, International Grasslands Congress, Moscow, Biological and Physiological Aspects of Intensification of Grasslands. Utilization: 325-332.

Teitzel, J.K., Abbott, R.A. and Mellor, W. 1974. Beef cattle pastures in the wet tropics. Queensland Agricultural Journal 100: 204-210.

Thomas, R. and Humphreys, L.R. 1970. Pasture improvement at Na Pheng, central Laos. Tropical Grasslands 4: 229-239.

Williams, C.H. and Andrew, C.S. 1970. Mineral nutrition of pastures. In "Australian Grasslands." (ed.) R.M. Moore. Australian National University Press, Canberra: 321-338.

7
A Multidisciplinary Approach to Tropical Pasture Improvement

W. T. Williams
R. L. Burt

INTRODUCTION

In earlier sections of this publication, there are examples of how the provision of well adapted pasture plants can be of major agronomic and economic significance. Stylosanthes hamata cv. Verano is one such plant; its ability to grow and fix nitrogen in dry tropical areas makes it possible to utilize huge areas of land which were previously of very little value. Few people realize, however, that S. hamata had been introduced into Australia many years previously, in 1944. This earlier introduction had excited little interest, presumably because its performance had been unsatisfactory. With the benefit of hindsight, it is believed that this failure was predictable. The plant concerned had been introduced from a subtropical zone and one in which it is likely to have been found on alkaline soils. It was tested on acid soils in more tropical regions; almost certainly it would have failed to nodulate satisfactorily under such conditions.

This is one of numerous examples which can be quoted to illustrate the need for a multi-disciplinary approach to the rational utilization of genetic resources. Had researchers appreciated the significance of the climatic and edaphic origin of the plant concerned, interest in S. hamata could have been aroused over 30 years ago. The same is true for many other species. One may appreciate, however, that it is logistically impossible to investigate every aspect of every introduction. Even a relatively simple operation, for instance the screening of collections for their rhizobial response patterns, would demand resources which are simply not available. Nor can it be expected that all relevant research will be carried out in one institute, or one country, or even at one time. Indeed there are many good reasons why this should not be so. Researchers are thus faced with a colossal dilemma: the need to produce information which will allow persons at present unknown, in disciplines with which the individual may not be familiar, to cooperate and coordinate with others.

The solution is that adopted by the taxonomists and described by Fosberg (1972). The investigator must attempt to use the information available to produce a biological framework on which in-

vestigators can 'hang' relevant information from elsewhere.   In
this way, all relevant information can be coordinated and used to
give a better understanding of the plants concerned. This body of
knowledge can then sensibly be used to choose strategic areas of
research in which the limited resources available can be most
gainfully employed.
        Such an approach has two prerequisites. First, since the in-
vestigator is interested in plants adapted to a wide range of climates and soils, it is axiomatic that the initial data base refer
to a wide spectrum of plant material. Such a base can only be obtained from the early stages of work with plant collections, before the accessions which fail to impress the researcher in the
area in question are shelved. One might refer to this as "genetic
resource data". Secondly, one must ensure that the relevant information is made available in some suitable format.
        The complete processing of any set of genetic resources data
can involve any or all of six processes; of these six, three are
(or should be) obligatory, and three are optional. The obligatory
processes are collection, organization, and storage; the optional
processes are publication, retrieval, and analysis. This chapter
is primarily concerned with analysis, since it is this aspect
which has been most neglected in the past and is yet essential for
the provision of 'biological frameworks.' Moreover, computational
methods appropriate for the analysis of such data are mostly of
recent origin and are still not widely known. Nevertheless, the
optional processes depend for their success on the satisfactory
completion of their obligatory forerunners, a brief consideration
of which is therefore necessary.

THE OBLIGATORY PROCESSES

Collection

        The collection of genetic resource data differs from that in
the experimental sciences in that it is widely spread over space
and time, and usually involves a number of observers. Plant collection in the field is largely confined to the accumulation of
physical materials, such as seeds and soil samples, from which
pertinent data must later be extracted. It is true that limited
information concerning the distribution and morphology of plants
and soils can, and should, be collected in situ; the morphological
and agronomic data which together are the heart of a useful data
bank must be obtained later, from observations in quarantine
glasshouses or field experiments. Mineral content of soils must
be determined in a laboratory, not necessarily that which has
sponsored the expedition; this is equally true of attributes of
specialized interest, such as rhizobial requirements of legumes,
or phytochemical data such as quinones, terpenes, or seed proteins. Information on disease resistance, if it is obtained at
all, is likely to appear much later, after an accession has been
widely grown. A particularly interesting case is that of climate. Broad climatic information can be, and often is, obtained

from published meteorological records prior to the expedition; but most collecting areas are poorly supplied with weather-stations, and the possibility of the existence of local unpublished records must always be kept in mind.

Perhaps the most important feature of collection is that it should avoid bias. Collecting expeditions, involving as they do, trained personnel and considerable expense, are not undertaken lightly. A collector may himself only be interested in plants for a specialized or restricted environment; but this does not absolve him from collecting materials and data which might be of value to others.

Organizations

It is all too easy for the main mass of morphological-agronomic data to be held in one laboratory, and the more specialized descriptors to be recorded in one or more laboratories elsewhere. A worker new to the field may then be unaware that subsidiary data sets exist. Ideally, all data concerning a single accession should be together in one place. The world data banks for major crop plants already approach this ideal; but for forage plants the situation is far less satisfactory, with many scattered and unorganized data sets.

Storage

Early plant-collecting records were stored in manuscript. Manuscript tables, however, are easily lost or mislaid, difficult to retrieve, and with the passing of years may become indecipherable. Edge-punched cards had a brief vogue, but the mechanical difficulties of handling them are so great that they are now seldom used. For a considerable time, therefore, center-punched 80-column cards were used, and many data banks still exist only in this form. Such cards, however, unless stored under very stringent conditions, are far from immortal; a card misplaced or lost can be disastrous in retrieval or analysis. It seems likely, therefore, that storage in the future will be on magnetic tape, or disks, or such other storage media as may supersede even these. It is then essential that the data be fully documented; even the most carefully prepared magnetic tape is useless if the nature of the data it carries is unknown.

THE OPTIONAL PROCESSES

Publication

If data are not published in some form, a worker new to the field cannot know that they exist, and they might as well not have been recorded. For small data sets conventional publication on paper is practicable, as has been done for the CSIRO Stylosanthes collection. For world-scale data sets this type of publication is hardly practicable, though it has been nearly achieved for major crop plants such as <u>Vigna</u>. However, data storage on a medium such as magnetic tape can be regarded as publication if copies of the tape can be made accessible to interested workers. All that is then necessary is to ensure that the nature of the contents of the tape is published in conventional form so that its existence is widely known.

Retrieval

We do not propose to deal with retrieval here. Retrieval always represents an attempt to obtain the answer to a specific question, a matter of immediate importance to the enquirer. Computational systems for data base management and efficient retrieval have reached a high degree of efficiency, although difficulties still occur because of the existence of rival systems.

Analysis

By 'analysis' is meant the elucidation of some simpler structure or pattern in a large and visually intractable set of data. It is not concerned, as is retrieval, with the answer to specific questions; nor is it concerned, as is statistical analysis, with testing a specific hypothesis. Its purpose is to <u>generate</u> hypotheses which might later be amenable to test, or to suggest specific questions which it might be profitable to answer. Its own answers are not unique; as MacKay (1969) has pointed out, no set of data possesses an inherently unique pattern. There is only what he has called a pattern-for-an-agent, a pattern which is profitable for a particular user. Fortunately, most workers in the field of genetic resources have substantially similar interests; as a result, so-called 'general purpose' elucidatory programs have proved widely useful. The major current difficulty concerning such programs is that the most powerful of them are still only practicable for relatively small data sets. Many of them are computationally impracticable or prohibitively expensive for more than a few hundred accessions; however, with the ever-increasing speed of modern computers, we may reasonably expect that this restriction will gradually be lifted.

There is a further difficulty. A decade ago there were few programs anywhere in the world suitable for this type of work; now there are several, most of which provide a number of alternative options. The newcomer's main difficulty now is to choose, from among the many 'pattern analysis' approaches available to him, that method most likely to be useful in his specific case. In the sections which follow, we summarize the main options available and provide examples illustrating the type of information they can provide.

THE METHODS OF ANALYSIS

The Basic Strategies

One may assume the existence of a data set representing a number of plant accessions (which in numerical works would be called 'individuals', 'elements' or 'operational taxonomic units') defined by one or more sets of descriptors (often called 'attributes'). The analytical approach can then be regarded as depending on two decisions; these are not mutually exclusive, since more than one analysis may be undertaken. For a single set of descriptors the first decision is whether the strategy is to be normal or inverse. In a 'normal' analysis it is some simpler configuration of the accessions which is sought; the descriptors do not appear in the results, except in so far as it may be possible to explore the relative importance of individual descriptors in defining the configuration obtained. In an 'inverse' analysis the data set is in effect transposed, and the configuration sought is that of the descriptors themselves. This choice presents no problems. The second decision is whether to resort to classification, which assumes the data set to discontinuous, or ordination, which assumes it to be continuous or largely so. This choice is both important and difficult, and requires further consideration.

Classification

The use of a classificatory computer program is equivalent to postulating that discontinuities exist in the data, which are to be identified as efficiently as possible. It must be realized that, even if the data set is substantially continuous, a classificatory strategy will 'make the best of a bad job' and find the best chopping points that it can. Moreover, many programs require the user to state in advance the number of groups he expects, and this is the number he will get. Algorithms[1] for assessing the optimum number of groups, and for ensuring that random data are not subdivided, are beginning to appear, but are not yet generally available.

Classifications are not unique, and the results obtained depend more on the numerical strategy employed than on the data themselves. Genetic resources data typically have mixed attribute types, which limits the number of alternative programs available, but even here a variety of strategies is available, a detailed discussion of which is beyond the scope of this chapter. The main disadvantage of classification is that it provides no information on intra-group structure--e.g., whether a given accession is central or peripheral in its group--and it provides little or no information concerning the relationship among groups. Nevertheless, it is often a convenient starting point and is still the only computationally practicable solution for very large data sets.

---

[1]Algorithms--A set of well-defined rules for the solution of a problem in a finite number of steps.

Ordination

There are now three main techniques of ordination. The first is principal co-ordinate analysis (Gower, 1966) which assumes that the data set could have been generated by the linear combination of a small number of underlying 'factors.' Unfortunately, genetic resource data usually contain discontinuities or non-linear relationships or both. The result is that the first few Gower vectors extract only an unacceptably small proportion of the variance in the data matrix. It has, therefore, largely been superseded by the minimum spanning tree (MST). This MST has the advantage that it provides a visual summary of all the individual accessions, though it gives little or no indication as to where the data are to be subdivided if this is considered desirable. Moreover, the information it displays is limited. If we consider three accessions in linear sequence on the tree A-B-C, the distances AB and BC are known, but AC is not, except by appeal to the original inter-accession distance matrix. Again, methods of overcoming this difficulty exist in the literature, but are computationally cumbersome and not easily available. Recently, therefore, a network technique has been devised, which uses both first and second, and sometimes higher, nearest neighbors. This has the great advantage that it displays a configuration consisting of islands of strongly-bonded continuous variation linked by weak bonds that represent discontinuities.

A Comparative Example of 'Normal' Analysis

In this section the results from three types of analysis, all on the same data set, are presented and compared so as to show the value and limitations of each. The current data set comes from a set of Stylosanthes accessions, newly introduced from Brazil and Venezuela. Little was known about the accessions concerned, or the species from which they came, and even the taxonomic status of some was in doubt. The first requirement was to obtain a synoptic view of the variation present and of the relationships among the various accessions. The reasons for setting these objectives have been detailed elsewhere (Burt et al., 1971). Included in these is the need to nominate duplicate introductions and to be able to communicate to others the 'type' of plant encountered (as shown by Krull and Borlaug (1970), failure to carry out the latter negates the value of plant introduction programs). Accordingly, the plants were grown in a spaced-plant field experiment, together with 'standard' accessions for comparative purposes, and various 'morphological' and 'agronomic' characteristics measured. The morphological information, readily available from pressed specimens, is not a usual requirement in this type of work and we shall return to its value at a later stage. The resultant data set was then subjected to three types of analysis.

The minimum spanning tree (MST). The results from the minimum spanning tree (MST) analysis are presented in Figure 7.1 (after Burt and Williams, 1979). Here the distance between neigh-

263

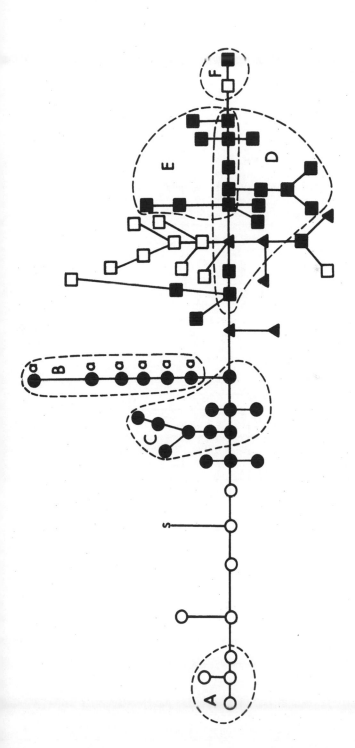

Figure 7.1. Minimum spanning tree for _Stylosanthes_. (Open circles, S. humilis; closed circles, S. hamata; solid triangles, S. sp. aff hamata, S. scabra; open squares, S. viscosa; S, S. subsericea. Zones A to E are considered further in Figure 7.2. For meaning of 'a', see text.)

boring accessions is a measure of their overall morphological-agronomic similarity. It is immediately obvious that the various species tended to separate from each other, even though no attempt had been made to include 'taxonomic' characteristics in the analysis; indeed 'key' attributes, such as the length of the axis rudiment, were not included. The species, moreover, were arranged in a sequence which is biologically meaningful; we pass from the annual S. humilis to the weakly perennial S. hamata to the strong perennials found in S. viscosa and S. scabra. Agronomically this represents a trend from plants with no ability to stay green in the dry season through to those which tend to do so. There are, however, no strong discontinuities between the species, S. hamata tending to overlap with S. humilis on one side and S. scabra on the other. Whereas the results tend to support the taxonomic delineations of the species, such distinctions are not as clear as one might wish. Nor can one accept that the high order distinction between S. hamata and S. humilis (placed in separate taxonomic subsections) is a useful guide to the overall similarities and agronomic characteristics of the species. It is relevant to note here that this may not be a typical situation. Where similar work has been carried out with other genera (Burt and Williams, 1979) there are clearer distinctions between the species.

It is now possible to 'hang' other information onto this framework, and this is illustrated primarily by reference to S. hamata. The S. hamata accessions marked 'a' on the figure are, as far as is known, all diploid, form effective rhizobial associations only with bacteria from alkaline soils and were collected from alkaline soils, usually in the Caribbean Islands. The only accession so far studied is daylength neutral. The remaining S. hamata accessions were tetraploids, which tended to form effective associations with rhizobia from both acid and alkaline soils and which were collected from acid soils. The one accession studied was a quantitative short-day plant. The two sets of accessions differed in morphological and agronomic characteristics; this will be docuumented subsequently. We now have a much better 'picture' of the plants concerned.

It is also of interest to consider the relationship between S. hamata and its neighboring species. On the left (Figure 7.1) we find that S. hamata 'merges' into S. humilis, which is diploid, found on acid soils in sub-coastal Venezuela, accepts rhizobia from acid soils and is an obligate short-day plant. It is plausible, therefore, that the tetraploid S. hamata, from coastal Venezuela, is a hybrid arising from Venezuelan S. humilis and Caribbean S. hamata (Date et al., 1979).

To the right of S. hamata is S. scabra (Figure 7.1). The species are linked by a plant designated S. sp. aff. hamata. This plant is a woody perennial, very similar to S. scabra, and quite unlike the herbaceous S. hamata; taxonomic identification has, however, placed it with the latter species. It is tetrapoid and occurs in sub-coastal Bazil; S. hamata, probably introduced from the Caribbean, is found on the coast. Again the possibility of hybridization cannot be discounted; certainly it occurs when previously isolated species are brought together (see, for instance,

Verdcourt, 1970).
This approach has raised a hypothesis which merits genetical
investigation, it has aided the sensible selection of this objective and defined the material which should be used in such an investigation. Moreover, from the results of isozyme[2] studies, it
would seem that this hypothesis is correct.

Network analysis. Another useful type of framework is that
provided by the network analysis. In Figure 7.2 the collection
splits into 12 weakly linked groups (weak, intermediate and strong
linkages being shown by dotted, single and double lines, respectively). The general configuration of the collection has remained
unaltered (compare Figure 7.1 with Figure 7.2) with S. humilis
being linked through S. hamata to S. scabra and so on. In many
cases the individual groups contain the same accessions, placed in
the same order and with the same accession noted as being most
similar (close spacing on the MST being equated with double bonds
on the network). It is, however, informative to consider briefly
the differences in configuration and groupings between the analyses.
Most of the differences in groupings were small and would
have little or no effect on the interpretation of the biology of
the system. One of the major differences concerns the widely
placed accession marked a' in group B of Figure 7.1. The network
analysis places this in a separate group (IV of Figure 7.2) but
even this major change simply re-emphasizes the distinctive nature
of accession a'. Major changes in configuration are also primarily concered with the three accessions placed in the latter group
(VI).
Our initial reaction is to distrust this grouping, largely
because it contains three named species. Examination of the raw
data, however, indicates that the analysis has really drawn to our
attention marked similarities which we would normally overlook.
The two tightly linked members of this group (placed close together on the MST, in group F in Figure 7.1) consist of an accession
of S. viscosa and one of S. scabra. They are quite different from
other accessions of these species; the S. viscosa accession, is,
for example, only slightly viscid, is late flowering and has relatively large seed, therefore resembling the S. scabra accession
and, in 22 of the 29 non-numerical characteristics available for
comparison, is identical. It is the remaining attributes, primarily those related to the very characteristic seed of S. viscosa,
which tend to make one intuitively separate these two accessions,
and places them in different species. The third member of the
group, the accession of S. hamata labelled 'a', is more weakly
linked to them than they are to each other. It shows many similarities with the diploid S. hamata accession of group V but differs from them in minor morphological characteristics and in its
late flowering habit and associated features such as late seeding;
the latter characteristic causes it to resemble the S. scabra accession. It is probably better placed with the other S. hamata

[2]Isozyme-an electrophoretically distinct form of an enzyme.

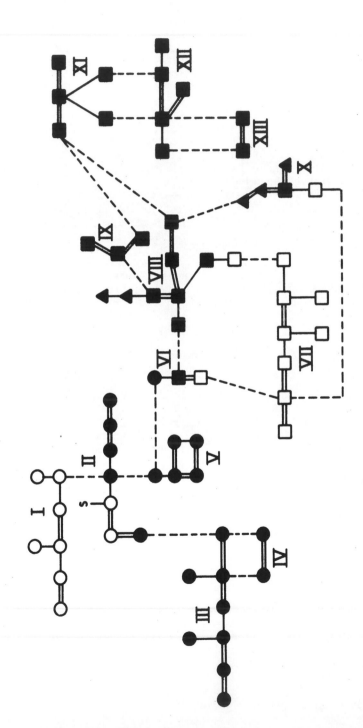

Figure 7.2. Network analysis of a Stylosanthes collection. (Open circles, S. humilis; closed circles, S. hamata; solid triangles, S. sp. aff hamata; solid squares, S. scabra; open squares, S. viscosa; S, S. subsericea. Weak, intermediate and strong bonds shown by dashed, single and double lines respectively.)

accessions, in the terminal position found on the MST.

The major changes in configuration are concerned with this trio of accession. In the MST their distinctive nature is recognized by their terminal, isolated placement. In the network, however, each individual of the group is used as a link to relevant species, S. hamata to S. hamata, S. scabra to S. scabra, and S. viscosa to S. viscosa. This does not imply that the group per se is central as it would on the MST.

It is decided to refrain from interpreting the biological significance of the network; clearly it closely resembles that obtained from the MST. The choice of the method to be used obviously depends upon the needs and preferences of the user. Further refinements of these methodologies are currently under investigation.

Classification. When classification is the first, or only, method of analysis to be used, it is usually necessary to nominate the number of groups to be obtained. For the current sample the choice was group 8 and the resultant hierarchy is presented in Figure 7.3. Although such hierarchies are assembled in an ascending fashion it is usual to read them from the top downwards as shown.

The first dichotomy (I) separates all accessions of S. humilis and all but one of the S. hamata's from the perennial species. From the associated program GROUPER we find that the former have higher dry matter yields (mean values of 454 and 165 g/plant for the two sides of the hierarchy); are poorly persistent; have hairier leaflets and pods with longer beaks (3.2 and 1.6 mm), etc. Next (II) most of the S. viscosa accessions are separated from the other perennials; they have a characteristically 'square' shaped seed, shorter main stems, stipules and pods, and individual seeds of low weight (1.01 and 1.81 mg). Now for (III), all but one of the S. humilis accessions are removed; they flower at shorter daylengths, have internode bristles, a high number of branches arising from the crown, etc. This process is continued, the agronomic information reinforcing the morphological. It is now possible to utilize such information to provide general purpose descriptions of the groups and these can be used to convey information to those interested in, but unfamiliar with, the collection. Agronomic data are of obvious interest here. Often, however, one should not neglect morphological information; it can be equally important. In Stylosanthes, seed color, a characteristic of little interest to most agronomists, is a valuable 'marker' characteristic for the plant breeder, and he must be made aware that it exists, in which accession it belongs and which characteristics are usually associated with it; other characteristics may be related to agronomic characters (hairy leaves may confer a 'lateral' resistance to insect-borne diseases) and are easily measured. Perhaps the best example to illustrate the value of morphological data concerns the cultivar Verano. Many initially regarded this plant as simply a vigorous accession of S. humilis; subsequently it was placed with other accessions of S. hamata. It was an atypical member of this group, however, in that it had elongate seeds, pale pods, and a

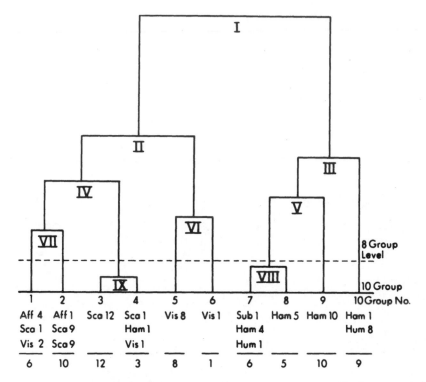

Figure 7.3. Classificatory hierarchy of a <u>Stylosanthes</u> collection at the 8 and 10 group level. (aff, <u>S</u>. <u>sp</u>. <u>aff</u> <u>hamata</u>; ham, <u>S</u>. <u>hamata</u>; hum, <u>S</u>. <u>humilis</u>; sca, <u>S</u>. <u>scabra</u>; sub, <u>S</u>. <u>subsericea</u>; vis, <u>S</u>. <u>viscosa</u>. Groups are numbered I to 10 and number and species content of each group is shown. Dichotomies are numbered I to IX.)

semi-erect habit (Burt <u>et al</u>., 1971). Without the morphological information it is quite possible that another member of this group would have replaced this accession in sward trials, performed poorly (as they did) and the potential of Verano would have gone unnoticed.

The groupings of the accessions closely resembled those obtained from the MST (the trio of the network group VI again being problematical). It was clear, however, that further division was necessary and dichotomies VIII and IX were made (Figure 7.3). The diploid <u>S</u>. <u>hamata</u> accessions were separated from the remainder (Group 7) and the network group VI members from the monospecific group 3. Group 2 probably also warrants subdivision (it occurs in two different zones on the MST). When used alone the absence of a 'stopping' rule can cause some problems as mentioned earlier. Another problem is associated with configuration; from the hierarchy we cannot describe for instance, whether the S. viscosa accessions in group 5 and 6 are most like the predominantly S. scabra accessions in groups 1 and 2 or those in 3 and 4. Nor can one determine the position of an accession within a group; it is not apparent, for instance, that one of the S. hamata accessions in group 10 (Figure 7.3) is peripherally placed ('a' of Figures 7.1 and 7.2). This seriously limits one's ability to biologically interpret continuously varying plant collections. If, however, this does not occur (evident from the MST or network analysis), then classification may still be appropriate.

Genetic resource data can be analyzed and presented in such a way as to provide a useful, workable framework on which data from other sources can sensibly be arranged.

THE USE OF DERIVED DATA

Although the basic data set may consist of accession x descriptors, further information may exist which enables the accessions to be grouped in some way, and the configuration of the resulting groups then explored by any of the previous methods. A single example is presented here and the reader can visualize other circumstances in which such methods could be utilized.

This example is concerned with work stimulated by that described previously. It is based on a collection of S. hamata which is believed to be representative of the species and the objective was to explore the relationships between populations from different geographical areas. As already seen, however, S. hamata is a 'wide' species with some forms being more similar to other species than they are to each other. Where such species occurred in areas from which S. hamata had been collected, therefore, representatives of these species were included in our collection. To this were added 'off-types,' plants which had arisen in evaluation or seed production plots, in which S. hamata had been grown. The entire collection comprised some 226 accessions, 171 of S. hamata and 4 of S. sp. aff. hamata, 26 of S. humilis, 4 of S. subsericea, 8 of S. sympodialis, 3 each of S. calcicola and S. scabra, 1 of S. mexicana and 6 'off-types' (including one known S. humilis x S. hamata hybrid. Morphological, agronomic and taxonomic data were also recorded.

It was first necessary to define the geographical areas which we wished to compare. Details of the procedure adopted are presented elsewhere (Williams et al., 1980b); briefly it involved the recognition of disjunct areas for S. hamata together with the wider floristic affinities noted by Howard (1973). Table 7.1 lists the areas recognized and the species found in each; Figure 7.4a shows their location. The similarities between the plant populations from these areas can now be investigated using a network analysis (Figure 7.4b.).

It is at once apparent that these areas have been split into 4 groups and that these are very largely geographically based (Figure 7.4a). Quite remarkably the Caribbean Islands have been divided into two zones, and each linked to either coastal South

TABLE 7.1
The Origin of <u>Stylosanthes</u> Species Used in the Experiment

Origin	Species	No.
1. Florida	S. hamata	23
	S. calcicola	1
2. Bahamas	S. hamata	1
3. Cuba	S. hamata	3
4. Dominican Republic	S. hamata	2
	S. humilis	1
5. Puerto Rico	S. hamata	1
6. Antigua	S. hamata	57
7. Berbuda	S. hamata	10
8. Nevis	S. hamata	15
9. Guadeloupe	S. hamata	2
10. St. Lucia	S. hamata	3
11. Margarita	S. hamata	3
12. Curacao	S. hamata	7
13. Paraguana	S. hamata	4
	S. humilis	6
14. Maracaibo	S. hamata	15
15. Baranquilla/Santa Marta	S. hamata	17
	S. humilis	7
16. Panama/Costa Rica/Honduras/	S. scabra[1]	1
Guatemala/Southern Mexico	S. mexicana	1
	S. subsericea	4
17. Barinas/Barquisimeto/Calobozo	S. hamata	4
	S. humilis	2
18. Ecuador	S. sympodialis	8
19. Australian "off types"		6
20. Australia	S. humilis	2
21. The Yucatan Peninsula	S. calcicola	2
	S. humilis	7
22. Brazil	S. hamata	3
	S. sp. aff. hamata	4
	S. scabra	3
23. Villavicencio	S. humilis	1

[1] Introduced as S. hamata but resembles S. scabra (see p. 25 of Edye et al., 1974) and S. sp. aff. hamata described previously (Burt and Williams, 1979).

America or southern U.S.A., in precisely the fashion (cf Howard, 1973). This would suggest, therefore, that variation within S. hamata is geographically based.

This is, not invariably so; the strong linkages between areas 1/6 (Florida and Antigua) and 20/33 (representing S. humilis accessions from naturalized Australian populations and Villavicencio) are somewhat unexpected. There are probably historical explanations for these links. Antigua, for instance, was the main British naval base in the region for many years following its establishment in the 17th century. Movement of troops, and presumably horses and fodder, between Florida and Antigua almost certainly occurred when the latter was under British control (1763-1783) or at a later date (1814) when British bases were established in Florida.

This network provides an elegant visual representation of the way in which variation is distributed within, primarily S. hamata. It is useful in that it allows those interested in the species to conceive the position of the material in which they are interested relative to the whole. Agriculturalists in Florida would not be unduly surprised by the failure of Verano (from area 14) in this region but may be interested in accessions from Antigua; those in Australia, in which the area 14 accessions have proved to be well adapted, and should clearly bias future testing towards accessions from area 15. Nor can one ignore the suggestion that the 'off-types' (area 18) have arisen from the related accessions from Brazil and Maracaibo.

Inverse Analyses

For many years inverse analyses of data containing mixed attribute types, especially if these included multistate nominals, were computationally intractable. The INVER algorithm of Lance and Williams (1979) now brings these within reach. Although inverse classifications have proved useful in such diverse fields as terrestrial ecology and social medicine, they have as yet found no use with genetic resource data. However, an inverse "minimum spanning tree" provides the possibility of ascertaining what relationships, if any, exist between morphological attributes and agronomic attributes such as dry matter yields.

Analyses of the grass genus Urochloa. To illustrate the application of this analytical method, we decided to work with information from two experiments concerned with the grass genus Urochloa. Suitable data from a leguminous genus are not readily available at this time. Like the Stylosanthes collection, the Urochloa collection was thought to be agronomically interesting but little or nothing was known about the accessions concerned or variation among or within the constituent species. Accordingly, two series of experiments were carried out. In the first, the collection was grown as spaced plants and agronomic and morphological data recorded; in the second, selected accessions were grown as small swards over a range of environments and various agronomic features noted.

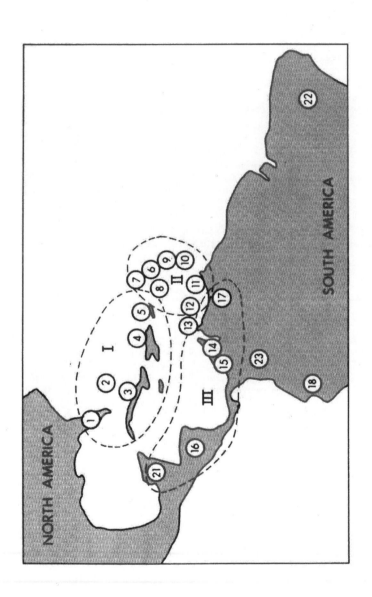

Figure 7.4a. Geographical areas from which the collection was obtained. (Areas 19 and 20 in Australia not illustrated. For meaning of Roman numerals, see text and Figure 7.4b.)

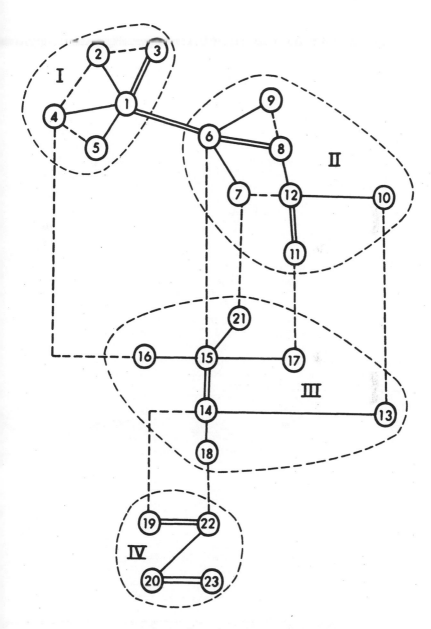

Figure 7.4b.   Network analysis of plant morphological data.
(Areas are shown in Figure 7.4a. Roman numerals
indicate regions nominated (see text). Dashed
lines, single lines and double lines indicate weak,
intermediary and strong bond strengths respectively).

Results from the inverse MST analyses of the first data-set are presented in Figure 7.5a (from Williams et al., 1980a). For the sake of brevity, consideration is given only those attributes on, or near to, the main axis and the discussion is further limited to their interrelationships with dry matter yield.

Yield labelled 1 in Figure 7.5a, is centrally placed in the MST; high yield is therefore associated with high values for the attribute groups, II, III and IV and to a lesser extent with those in V. Accessions with high values for Group I attributes are low yielding (the latter, hairiness of leaf surface, leaf margin and stem internodes and characteristic of the low yielding U. stolonifera accessions). Group II attributes all relate to rhizome/stolon development and accessions which show high values for these, together with long stems (III) are almost always from the species U. mosambicensis and U. pullulans. The distinction between these species is often not taxonomically accepted. Group IV attributes contain three which are coincident (two for regrowth after cutting and one for tiller number) and this is a relationship is expected. The remaining attribute in this group (number of flower heads) is also of necessity at least partially dependent upon tiller number. Finally, there are the Group V attributes. These are related to 'size', of leaves, stems, and inflorescence heads, and high values of these almost invariably denote U. bolbodes.

It seems, therefore, that under spaced plant conditions, the high yield can be obtained in two ways: (a) in U. mosambicensis, by selecting (or breeding) for rhizome development, long stems, and high rates of regrowth; and (b) in U. bolbodes, by selecting for 'big' characteristics, again with high regrowth ability. It is tempting to suggest that a combination of these characteristics might produce an even more superior plant. Such accessions do, in fact exist; one accession of U. bolbodes exhibited marked rhizome development and one of U. pullulans possessed several of the large size characteristics. Both, however, were of very low persistence.

It is generally known that it is difficult to select sensibly both objectives and parents for a breeding program. At least part of the problem occurs because little or nothing is known about which characteristics tend to be related. It is plausible to believe that the type of results presented could help to alleviate such difficulties.

In the second experiment, yield was measured under regularly cut sward conditions over a range of environments. Attribute 8, termed the 'yield ratio' is a measure of the ability of an accession to yield well over a range of environments. Such an ability is often not predictable from performance under spaced plant conditions. The second analysis was done to determine which attributes, as measured under spaced plant conditions, are most likely to predict success in the field. The analysis was simply re-run for a smaller set of plants and the yield ratio attribute was in-included in the data set.

To a very large extent, the configuration of the MST, and the positions of the various attributes remained unaltered. Only the relevant part of the MST has therefore been presented (Figure

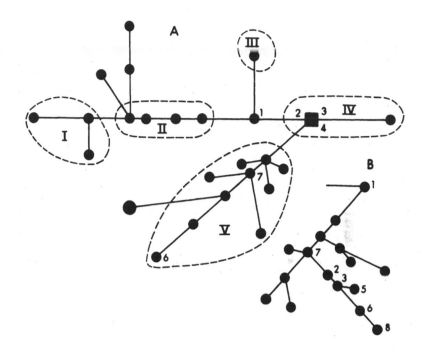

Figure 7.5a.   Inverse <u>Urochloa</u> tree.
  (1. dry matter yield;
   2. regrowth two weeks after cutting;
   3. regrowth three weeks after cutting;
   4. tiller number/plant;
   5. inflorescence number per plant;
   6. internode length;
   7. number of racemes in primary inflorescence;
   8. yield ratio.)

Figure 7.5b.   Inverse <u>Urochloa</u> tree for reduced set.   (Attribute
  numbering as in Figure 7.5a.)

7.5b).
     It is apparent that yield ratio (8) is only poorly related to
yield under spaced plant conditions (1).   It is much more closely
related to internode length (6), the two regrowth characteristics
(2 and 3), raceme number (7) and the number of flower heads/plant
(5).   In retrospect, this is reasonable; the rates of regrowth and
internode length reflect the ability to add dry matter after defoliation and the flowering characteristics would be related to seed
production and ability of the accession to "thicken up."   Seed
yielding ability should be measured in future spaced plant work.
     It would be logical to use the tactics employed here to coor-
dinate research effort; an institute with an interest in a given

genus could produce the basic morphological/agronomic data set and another could utilize this, for example, to find characteristics related to their measurements of disease resistance. If such an approach is to be successful, then the initial data set must be sufficiently broad to cope with such a demand. Internode length (6), a morphological attribute of little significance when one question was posed (factors related to yield--Figure 7.5a) was of major significance when the question was changed (factors related to yield ratio--Figure 7.5b).

Multiple Attribute-Sets

It is in this situation that the true multi-disciplinary case occurs. For example, a set of Stylosanthes accessions may be characterized by morphological-agronomic information, by rhizobial responses, by the climate and/or soils of the place of origin, or by phyto-chemical data such as seed proteins obtained from electrophoretic study. The problem is essentially canonical and is always inverse, in that what is required is some comparison of different attribute-sets over the same set of accessions. We can distinguish five possible approaches:

Simple comparison. The two sets are processed separately and the results compared, often visually; as an example, compare Figure 7.6 with Figure 7.4a. Figure 7.4a has been discussed previously; it is derived from morphological/agronomic information obtained from a set of Stylosanthes accessions and is used to explore the relationship between various geographical areas. Figure 7.6 deals with the same plant collection and has the same aim in view, but utilizes isozyme banding patterns rather than morphological/agronomic data (Robinson et al., 1980).

The results obtained from the two analyses are in substantial agreement. For example both separate the Caribbean Islands into two identical groups (I and II) and both show an unexpectedly strong affiliation between Florida and Antigua (areas 1 and 6). For some purposes at least, it may be possible to replace morphological/agronomic measurements with the more simply obtained isozyme measurements. There are, however, differences between the two sets of results: the putative hybrids (area 19) appear to be most similar to the S. hamata accessions from Maracaibo (area 14) where isozyme data are used but less so on the basis of morphological/agronomic information. It is usually assumed that isozyme data more closely reflect the genetic structure of the plants than morphological information, but further studies are warranted.

Canonical coordinate analysis. This, the most rigorous approach, has occasionally been used (Williams et al., 1973); owing to the difficulty of meeting the basic assumptions of principal coordinate analysis, it is seldom successful.

Single-attribute superimposition. If one takes a set of accessions defined by, e.g. climatic data, a principal coordinate analysis of the data is usually possible. The points on the resulting graph strictly represent the areas from which the accessions were collected but if they are labelled with their taxonomic names of the accessions, we are in effect regarding the taxonomic name as a single multistate nominal attribute, and superimposing this attribute on the principal coordinate analysis.

In the absence of a system of climatic classification suited to plant introduction purposes, this method was employed to explore the relationships between the climates in which plants were collected and the species found and their performance in different climates in Australia. The ordination (Figure 7.7) revealed that the major variation between the sites was primarily in temperature and rainfall (Vector I); high positive values were associated with high temperatures and low annual rainfall and high negative values with the low temperature and high annual rainfall. The second vector (II) was associated with, at the positive end, high altitudes and relatively short growing seasons. The climates concerned were very variable; site 1 had a mean monthly maximum temperature of 34°C, a minimum of 25°C and an annual rainfall of 387 mm; for site 2 the equivalent values were 29°C and 12°C and 3969 mm. For convenience, the climatic continuum has been arbitrarily divided into 4 zones (Figure 7.7).

Some species, such as S. fruticosa (from Africa) were found over a wide range of climates. Other species appeared to be much more restricted in their distribution, such as S. humilis to drier zones and S. guianensis to wetter areas. Such observations would successfully predict the known adaptational patterns of these species when introduced elsewhere. It is, moreover, possible to be more specific. It is found, for instance, that the legumes which have been most successful in the dry Zone B climates (e.g. S. hamata cv. Verano) were collected from the semiarid environments of Zone A environments; those successful in Zone C (where South Johnstone is placed), where ability to grow into the cooler, drier season was sought, came from similar environments but ones which experienced cooler, drier seasons. It appears, therefore, that the legumes are most agronomically desirable for a given area are likely to be found in similar but somewhat more stringent environments. Nor is this relationship restricted to Stylosanthes (Burt, Reid, and Williams, 1975).

This type of analysis does not require detailed information about climates or plant performance and seems to be a useful approach in coordinating collecting and evaluating data. The information could be used to select areas for future plant collecting or in choosing suitable evaluation sites for existing collections. The same method has been used to study plant distribution in relation to soil mineral characteristics and rhizobial response patterns in relation to species and climate. It is generally useful in coordinating information from different disciplines.

Classificatory superimposition. Assume that two data sets, A and B, represent the same set of accessions defined by two sets of descriptors. One can subject A to a hierarchical agglomerative classification and then impose the hierarchy obtained from A on the data from B. The disposition of the attributes of B on the

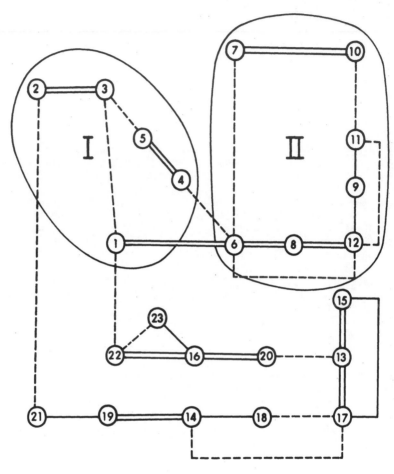

Figure 7.6.   Network analysis of a <u>Stylosanthes</u> collection based
on isozyme banding patterns only. (For notation and
symbols see Figure 7.4b and Table 7.1.)

classification of A can then be examined by a 'diagnostic' program
such as GROUPER (Lance <u>et al</u>., 1968). Alternatively, because the
data from B have not been used in the classification of A, differences between the B attributes at any level of the A fusions can
be tested statistically.   This approach is illustrated, as follows: The classification was used to place <u>Stylosanthes</u> accessions into "morphological/agronomic groups"; the objective was to
know if these groups tended to be found in certain climates or locations.   The 'B' attributes nominated, therefore, contained two
locational attributes (1 and 2, latitude and altitude), two for
rainfall (3 and 8, length of wet season and total annual rainfall)

279

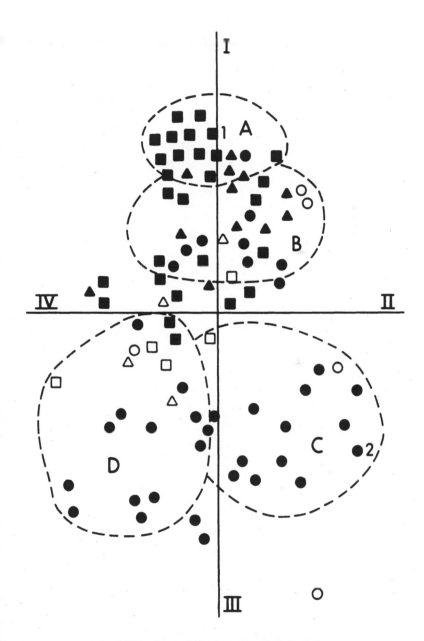

Figure 7.7.   Climatic ordination of a <u>Stylosanthes</u> collection.
(Solid circles, <u>S</u>. <u>guianensis</u>; open circles, <u>S</u>. <u>fru-
ticosa</u>; solid squares, <u>S</u>. <u>hamata</u>; open squares, <u>S</u>.
<u>sp</u>. <u>aff</u> <u>hamata</u>; solid triangles, <u>S</u>. <u>humilis</u>; open
triangles, <u>S</u>. <u>scabra</u>.)

and four (4 to 7) for maximum and minimum temperatures in the wet and dry seasons. The results displayed (Figures 7.8a and 7.8b) refer primarily to groups of S. guianensis (Figure 7.8a) and S. scabra (Figure 7.8b) (Burt, 1976).

In S. guianensis groups 5 and 6 are from higher latitudes, lower rainfalls and lower temperature regimes than 7 and 8; 8 comes from higher altitudes than 7, and 6 from areas with low night temperatures in the growing season (attribute 5). In S. scabra (Figure 7.8b) a more subtropical form (19) exists but this time from relatively high rainfall areas. There were no differences between the climates from which groups 17 and 18 originated, possibly because they were represented by few accessions and few data were available for statistical comparison. The value of this information has been discussed elsewhere. We simply note here that this method could be easily applied to other multi-attribute sets.

Inverse networks or trees. This process here consolidates the attribute-sets, and obtains a single data-matrix defined by a double or triple set of attributes. The relationships between the latter can then be explored by any of the above methods. Two examples are presented.

The first of these concerns a problem inherent in chemotaxonomic studies. Although it is possible to measure differences between plants, it is difficult to envisage the significance of such differences; frequently chemotaxonomic data have been regarded as secondary to other information. For this, and other reasons, it is desirable to relate such information to more readily observable features. One may attempt to do so for the morphological/agronomic and isozyme data from the Stylosanthes collection described previously

The approach adopted was simply to bulk the two data-sets and subject them to an inverse network analysis. The attributes were grouped into 10 regions; some contained only morphological or isozyme bands and other contained mixtures of both. Examples of these 3 types are presented in Figure 7.9 (taken from Robinson et al., 1980).

The first region (A) contained both types of attributes. Beak/pod ratio (S6) was strongly linked to beak length (S3), and both are strongly linked with the presence of the isozyme band 25. They are in turn linked to the presence of a shoulder on the seed (S1). The presence of a shoulder on the seed, a beak length of less than 2 mm or the presence of band 25 can be used to separate accessions of S. scabra, S. sp. aff. hamata, and S. sympodialis from all but three of the remaining accessions. These latter accessions, one from southern Brazil and two from Florida, have been identified as S. hamata but are quite atypical of this species. They are from the extremes of the geographical range of this species and bear morphological resemblance to S. scabra. Their taxonomic provenance merits investigations.

The next region (B) contains only morphological attributes; all are floral. Included in this region is F7, length of the axis rudiment, the presence or absence of which is used to define the

281

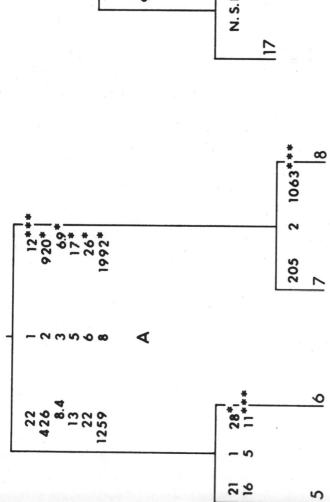

Figure 7.8a. Differences in the climatic backgrounds of various forms of *S. guianensis*.

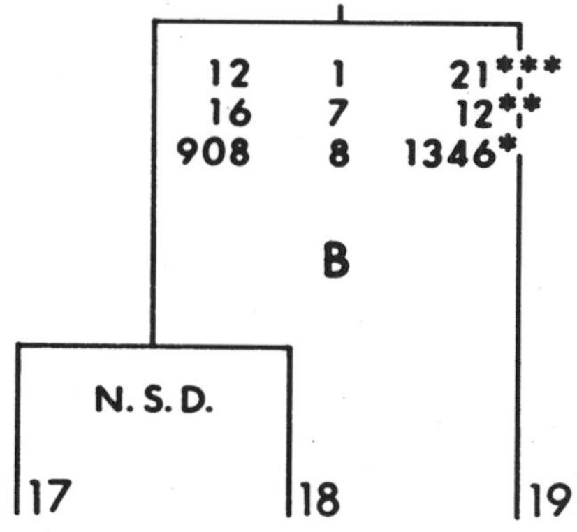

Figure 7.8b. Differences in the climatic backgrounds of various forms of *S. scabra*.

282

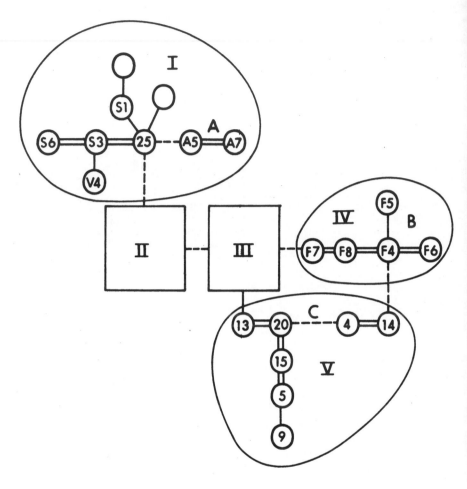

Figure 7.9. Inverse network analysis of a <u>Stylosanthes</u> collection; both agronomic/morphological and isozyme banding patterns are included. (Contents of groups II and III are not illustrated.)

(1) <u>Centrosema virginianum</u>. Edaphically, this species seems to prefer high available P levels (bond 1a). The forms contained in this collection require a long growing season (bond 1b). Both suggestions have been validated in field experiments (unpub. data).

(2) <u>Arachis sp</u>. Dominance was associated with high Ca levels in the soil (bond 2a). Reference to the literature suggests that this is likely to be so; the genus is often deemed to have a high Ca requirement and it is known (for other genera and species) that

two subgenera within Stylosanthes. Accessions with a long axis rudiment, the presence or absence of which is used to define the two subgenera within Stylosanthes. Accessions with a long axis rudiment and long inflorescences (F4) are S. sympodialis (from Ecuador) and the S. hamata accessions (excluding the two exceptions noted previously) from Florida.

Finally, there is region (C) which contains only isozyme bands. The accessions which exhibit these bands come from a variety of geographical locations and species and show no consistency in agronomic or morphlogical characteristics. At the present time presence of these bands can only be used to indicate genetic variation of an uncertain nature.

This analysis has helped to select those bands, or groups of bands, that can be used to delineate certain species and species forms; it has helped to reinforce the distinction between the two 'odd' S. hamata accessions from Florida and the remainder from that area. It has done little to consolidate the position of the various floral attributes as key taxonomic characteristics.

The second example concerns a set of accessions collected from Brazil, Antigua (West Indies), Australia, and Colombia. For each of 35 collection sites, one could nominate the quantity of the various leguminous species found and provide information on the soils and climates. It was necessary to ascertain which of the environmental factors were most closely associated with the presence of large quantities of the various species (Burt et al., 1979b).

The first step was to compute the separate inverse networks for the 19 floristic attributes (species) and for the 16 soil/climate attributes (Figure 7.10). The soil attributes were then masked out and the floristic/climate network computed; cross linkages were imposed on the basic network. The procedure was then repeated to provide cross linkages with the soil attributes. The floristic network contained five groups of species, details of which are presented elsewhere. Briefly, I is characteristic of legume associations found on red earth soils in dry tropical regions of Australia; II is of species frequently found on clay soils in the tropics; III contains species found in dry tropical environments on lighter soils in N.E. Brazil; IV is of legumes from drier parts of Brazil. Clearly these groupings are biologically real and could provide useful information.

The climatic/soil attributes also separated into weakly linked groups. Group A contained the four temperature attributes, weakly linked to soil depth. Group B contained three attributes, two for rainfall parameters and the third for soil N percent; in these soils high soil N was associated with high rainfall and long growing seasons. Finally, in Group C, high clay contents are associated with high levels of exchangeable Ca, Mg and K and high levels of extractable P; pH is linked to Ca. Such relationships are logical.

Cross linkages between the two networks can now be used to suggest which of the environmental factors are most closely associated with the dominance of the various species. The results from three species are illustrated.

284

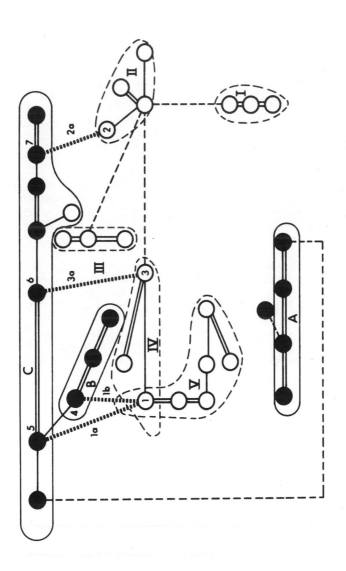

Figure 7.10. Network analysis of floristic and environmental information. (Groups A, B, and C are for environmental attributes and I to IV for floristic. 1. C. virginianum; 2. Arachis spp; 3. Teramnus labialis; 4. Number of months receiving 85-90% of the mean annual rainfall; 5. extractable soil P (ppm); 6. soil Mg (me per 100 g); 7. exchangeable soil Ca (me/100g).

high soil Ca levels stimulate the production of geocarpic seed and thus ensure survival under heavy grazing pressures. There were no strong linkages with any of the climatic attributes.

(3) Teramnus labialis. This species was most frequent when soil Mg levels were high (bond 3a) and where growing seasons were relatively long. Field studies have validated the latter (unpub. data) but the relationship with Mg remains unexplored; circumstantial evidence, however, would suggest that it is real.

It is logistically impossible to screen every accession or group of accessions for either mineral or climatic requirements. It is possible, however, that hypothesis generating analyses such as these can be used to indicate where such studies are most likely to be profitable.

CONCLUSION

Throughout this discussion, the various authors have indicated, not only the value of existing legume cultivars, but also the very real need to develop our genetic resources further. Should this happen, then an ad hoc approach to their collection, evaluation, and utilization will inevitably lead to inefficiencies, duplication of effort and the 'shelving' of material, useful in areas other than that in which it is held, will occur. This is wasteful of a scant resource.

In this section, the argument has necessarily stressed the need for data analysis and integration and their role in a hypothesis generating situation. Such hypotheses have been generated in the fields of genetics, plant geography, microbiology, taxonomy, and agronomy; several have subsequently been validated. In this way it may be possible to highlight strategic areas of research for those who wish to participate in the overall program. It is also relevant that most of the information used is that which is usually obtained from plant collecting, introduction, and evaluation programs, indeed much of it is regarded as "routine." Like Krull and Borlaug (1970) it is valid to submit that "The natural variability in collections has been ignored" and that "The major hurdle to unlocking their secrets (plant collections) and utilizing the valuable characters (contained) has been the investigator's inability to satisfactorily classify this variability".

REFERENCES

Burt, R.L., Edye, L.A., Williams, W.T., Grof, B. and Nicholson, C.H.L. 1971. Numerical analysis of variation patterns in the genus Stylosanthes as an aid to plant introduction and assessment. Australian Journal of Agricultural Research 22: 737-757.

Burt, R.L., Reid, R., and Williams, W.T. 1975. Exploration for, and utilization of, collections of tropical pasture legumes.

I. The relationship between agronomic performance and climate of origin of introduced Stylosanthes spp. Agro-Ecosystems 2: 293-307.

Burt, R.L. 1976. Case 3.3. The climatic background of Stylosanthes accessions. In "Pattern Analysis in Agricultural Science." (ed.) W.T. Williams. CSIRO and Elsevier Scientific Publishing Co. Melbourne and Amsterdam. pp. 164.

Burt, R.L. and Williams W.T. 1979. Strategy of evaluation of a collection of tropical herbaceous legumes from Brazil and Venezuela III. The use of ordination techniques in evaluation. Agro-Ecosystems 5: 135-146.

Burt, R.L., Pengelly, B.C., and Williams, W.T. 1980. Network analysis of genetic resource data III. The elucidation of plant/soil/climate relationships. Agro-Ecosystems 6: 119-127.

Date, R.A., Burt, R.L., and Williams, W.T. 1979. Affinities between various Stylosanthes species as shown by rhizobial, soil pH and geographic relationships. Agro-Ecosystems 5: 57-67.

Fosberg, F.R. 1972. The value of systematics in the environmental crisis. Taxon 21: 631-634.

Gower, J.C. 1966. Some distance properties of latent root and vector methods used in multivariate analysis. Biometrika 53: 325-338.

Howard, R.A. 1973. The vegetation of the Antilles. In "Vegetation and Vegetational History of Northern Latin America" (ed.) A. Graham Elsevier, Amsterdam, London and New York. pp. 1-38.

Krull, C.F. and Borlaug, N.E. 1970. The utilization of collections in plant breeding and production., pp. 427-440. In 'Genetic Resources in Plants" (eds.) O.H. Frankel and E. Bennett, Blackwell Scientific Publications, Oxford and Edinburgh.

Lance, G.N. and Williams, W.T. 1979. INVER: a program for the computation of distance-measures between attributes of mixed types. Australian Computer Journal 11: 27-28.

Lance, G.N., Milne, P.W., and Williams, W.T. 1968. Mixed-data classificatory programs III Diagnostic systems. Australian Computer Journal 1: 178-181.

Mackay, D.M. 1969. Recognition and action. In "Methodologies of Pattern Recognition" (ed.) S. Watanabe. Academic Press, London. pp. 409-416.

Robinson, P.J., Burt, R.L., and Williams, W.T. 1980. Network analysis of genetic resource data II. The use of isozyme data in eludicating geographic relationships. Agro-Ecosystems 6:111-118.

Verdcourt, B. 1970. Studies in the Leguminosae-Papilionoidae for the Flora of Tropical East Africa. I. Kew Bulletin 24: 1-70.

Williams, W.T., Edye, L.A., Burt, R.L., and Grof, B. 1973. The use of ordination techniques in the preliminary evaluation of Stylosanthes accessions. Australian Journal of Agricultural Research 24: 715-731.

Williams, W.T., Burt, R.L., and Lance, G.N. 1980a. A method for establishing character interpretations in plant collections, and possible applications to plant improvement programmes. *Euphytica* 29: 625-633.

Williams, W.T., Burt, R.L., Pengelly, B.C., and Robinson, P.J. 1980b. Network analysis of genetic resource data I. Geographical relationships. *Agro-Ecosystems* 6: 99-109.

Scientific Name Index

Ademesia 223
Aeschynomene 55, 223
Amnemus
  quadrituberculatus 111
  superciliaris 111
Andropogon gayanus 47, 91

Bothriochloa insculpta 47
Brachiaria
  brizantha 47
  decumbens 30, 33, 41, 47,
    79, 120, 163, 206
  humidicola 79
  mutica 29, 33, 41, 47, 79
  ruziziensis 47

Cajunus 223
Calopogonium 55, 223
  mucunoides 30, 48, 76
Canavalia 223
Cassia 55
Catenaria 99
Cenchrus
  ciliaris 47, 79, 87
  setigerus 47
Centrosema 18, 55, 56, 61-68,
  69-96, 183-184, 189, 192,
  193, 214, 221, 222, 223,
  232, 238
  acutifolium 91
  angustifolium 71, 74, 75, 91
  arenarium 71, 73, 75
  bracteosum 75
  brasilianum 70, 71, 73, 74,
    75, 89, 91
  capitatum 91
  grandiflorum 71, 74, 75, 91

Centrosema cont.
  haitiense 89
  kermesi 89
  lobatum 89
  macranthum 89
  macrocarpum 71, 75, 91
  magnificum 91
  pascuorum 70, 71, 72, 73,
    74, 75, 83-86, 91
  plumieri 71, 75, 89-90, 91,
    183, 223
  pubescens 30, 33, 41, 48,
    64, 69, 71, 72, 73, 74,
    75-82, 87, 89, 90, 91,
    92, 120, 125, 183, 190,
    199, 201, 202, 206, 214,
    223, 234, 239
  rotundifolium 71, 73, 75
  saggitatum 71, 75
  schiedeanum 75
  schottii 71, 74, 88-89
  venusum 75, 91
  vexillatum 74
  virginianum 70, 71, 72, 74,
    75, 86-88, 91, 183, 223,
    282, 284
  virginianum x brasilianum
    223
  sp. aff. acutifolium 71, 75
  sp. aff. pubescens 30, 71,
    82-83, 90, 91
  spp. 87
Centrosema mosaic virus 90
Cercospora 78, 82
Chapmannia 147
Chloris gayana 47, 79
Clitoria 69, 223

Printed and bound by CPI Group (UK) Ltd, Croydon, CR0 4YY

23/10/2024

01778244-0003